MARTIN BARRALL & DAVE PARRY

Hodder Arnold
A MEMBER OF THE HODDER HEADLINE GROUP

Orders: please contact Bookpoint Ltd, 130 Milton Park, Abingdon, Oxon OX14 4SB. Telephone: (44) 01235 827720. Fax: (44) 01235 400454. Lines are open from 9.00–6.00, Monday to Saturday, with a 24 hour message answering service. You can also order through our website www.hodderheadline.co.uk.

British Library Cataloguing in Publication Data
A catalogue record for this title is available from the British Library

ISBN-10: 0340 91530 7
ISBN-13: 978 0340 91530 1

First Published 2006
Impression number 10 9 8 7 6 5 4 3 2 1
Year 2012 2011 2010 2009 2008 2007 2006

This high quality material is endorsed by Edexcel and has been through a rigorous quality assurance programme to ensure that it is a suitable companion to the specification for both learners and teachers. This does not mean that its contents will be used verbatim when setting examinations nor is it to be read as being the official specification – a copy of which is available at www.edexcel.org.uk

Typeset by Pantek Arts Ltd, Maidstone, Kent
Illustrations by Mike Parsons at Barking Dog Art.
Printed in Great Britain for Hodder Arnold, an imprint of Hodder Education,
a division of Hodder Headline Plc, 338 Euston Road, London NW1 3BH by Bath Press Ltd

Contents

ELECTRONIC END USER SINGLE USE LICENCE AGREEMENT

FOR DiDA Unit 2: Multimedia CD-ROM Student Version software published by Hodder and Stoughton Limited (HS) under its Hodder Arnold imprint.

NOTICE TO USER:
THIS IS A CONTRACT. BY INSTALLING THIS SOFTWARE YOU AND OTHERS TO WHOM YOU ALLOW ACCESS TO THE SOFTWARE ACCEPT ALL THE TERMS AND CONDITIONS OF THIS AGREEMENT.

This End User Single Use Licence Agreement accompanies the DiDA Unit 2: Multimedia CD-ROM Student Version software (the Software) and shall also apply to any upgrades, modified versions or updates of the Software licensed to you by HS. Please read this Agreement carefully. Upon installing this software you will be asked to accept this Agreement and continue to install or, if you do not wish to accept this Agreement, to decline this Agreement, in which case you will not be able to use the Software.

Upon your acceptance of this Agreement, HS grants to you a non-exclusive, non-transferable licence to install, run and use the Software, subject to the following:

1. Use of the Software. You may only install a single copy of the Software onto the hard disk or other storage device of only one computer. If the computer is linked to a local area network then it must be installed in such a way so that the Software cannot be accessed by other computers on the same network. You may make a single back-up copy of the Software (which must be deleted or destroyed on expiry or termination of this Agreement). Except for that single back-up copy, you may not make or distribute any copies of the Software, or use it in any way not specified in this Agreement.

2. Copyright. The Software is owned by HS and its authors and suppliers, and is protected by Copyright Law. Except as stated above, this Agreement does not grant you any intellectual property rights in the Software or in the contents of DiDA Unit 2: Multimedia CD-ROM as sold. All moral rights of artists and all other contributors to the Software are hereby asserted.

3. Restrictions. You assume full responsibility for the use of the Software and agree to use the Software legally and responsibly. You agree that you or any other person within or acting on behalf of the purchasing institution shall NOT: use or copy the Software otherwise than as specified in clause 1; transfer, distribute, rent, loan, lease, sub-lease or otherwise deal in the Software or any part of it; alter, adapt, merge, modify or translate the whole or any part of the Software for any purpose; or permit the whole or any part of the Software to be combined with or incorporated in any other product or program. You agree not to reverse engineer, decompile, disassemble or otherwise attempt to discover the source code of the Software. You may not alter or modify the installer program or any other part of the Software or create a new installer for the Software.

4. No Warranty. The Software is being delivered to you AS IS and HS makes no warranty as to its use or performance except that the Software will perform substantially as specified. HS AND ITS AUTHORS AND SUPPLIERS DO NOT AND CANNOT GIVE ANY WARRANTY REGARDING THE PERFORMANCE OR RESULTS YOU MAY OBTAIN BY USING THE SOFTWARE OR ACCOMPANYING OR DERIVED DOCUMENTATION. HS AND ITS AUTHORS AND SUPPLIERS MAKE NO WARRANTIES, EXPRESS OR IMPLIED, AS TO NON-INFRINGEMENT OF THIRD PARTY RIGHTS, THE CONTENT OF THE SOFTWARE, MERCHANTABILITY, OR FITNESS FOR ANY PARTICULAR PURPOSE. IN NO EVENT WILL HS OR ITS AUTHORS OR SUPPLIERS BE LIABLE TO YOU FOR ANY CONSEQUENTIAL, INCIDENTAL, SPECIAL OR OTHER DAMAGES, OR FOR ANY CLAIM BY ANY THIRD PARTY (INCLUDING PERSONS WITH WHOM YOU HAVE USED THE SOFTWARE TO PROVIDE LEARNING SUPPORT) ARISING OUT OF YOUR INSTALLATION OR USE OF THE SOFTWARE.

5. Entire liability. HS's entire liability, and your sole remedy for a breach of the warranty given under Clause 4, is (a) the replacement of the Software not meeting the above limited warranty and which is returned by you within 90 days of purchase; or (b) if HS or its distributors are unable to deliver a replacement copy of the Software you may terminate this Agreement by returning the Software within 90 days of purchase and your money will be refunded. All other liabilities of HS including, without limitation, indirect, consequential and economic loss and loss of profits, together with all warranties, are hereby excluded to the fullest extent permitted by law.

6. Governing Law and General Provisions. This Agreement shall be governed by the laws of England and any actions arising shall be brought before the courts of England. If any part of this Agreement is found void and unenforceable, it will not affect the validity of the balance of the Agreement, which shall remain wholly valid and enforceable according to its terms. All rights not specifically licensed to you under this Agreement are reserved to HS. This Agreement shall automatically terminate upon failure by you to comply with its terms. This Agreement is the entire and only agreement between the parties relating to its subject matter. It supersedes any and all previous agreements and understandings (whether written or oral) relating to its subject matter and may only be amended in writing, signed on behalf of both parties.

The enclosed single-user licence CD-ROM contains various resources designed to assist your throughout your course. The CD-ROM has several features, including sections that correspond to the chapters in this textbook. These features include:

- Exercises: there are one or two exercises in this section for each chapter of the textbook. Each one will help you to consolidate your knowledge of a particular topic. Some require you to refer to a specific website, for others there are associated files that you can use to complete the exercise. Whenever associated files are mentioned in an exercise they are written in CAPITALS and a link to the file itself will appear in the 'Resource File' panel on the right hand side of the page.
- Tutorials: These demonstrations of how to use various software packages are available in those chapters (11-16) that relate directly to software applications. Short descriptions of each skill will appear when you roll the mouse pointer over each tutorial link.
- Useful links: this section contains useful links to other websites and resources relevant to each particular topic. If you are connected to the internet, simply clicking on each link will open the website in a new web browser window. Please note: All weblinks contain live content and information, and as such this content is liable to change. Hodder & Stoughton Ltd are not responsible for the content of any external website. An assessment of the appropriateness of each website for the intended audience should be undertaken by a responsible adult
- Resource bank: This area provides a range of dummy files for you to practice your software skills on. They are nothing more than examples of each type of file, to help you to get used to the sorts of files you will be working with when studying DiDA Unit 2. You can take each file and do whatever you want with it – stretch it, squash it, edit it, delete it...! Some of these files are duplicates of those that are used in exercises from the student and teacher sections; other files only appear in this section.

Before you start using this CD ROM, it may be helpful to refer to the 'How to use this CD-ROM' section, which you can access from the main menu screen. Included in this area is a tutorial that will provide you with a step-by-step introduction to accessing and using the different types of resources available on this CD-ROM.

Acknowledgements

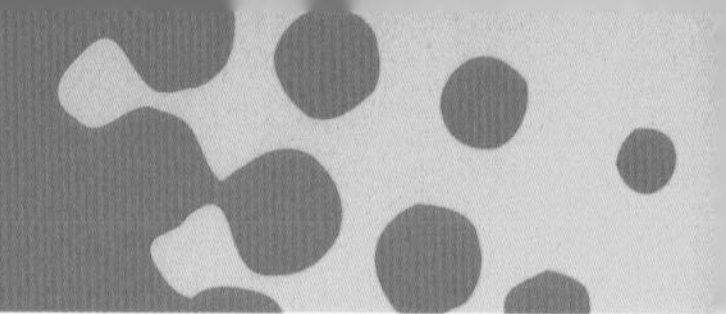

The authors would like to thank Drew Buddie for all his help in preparing the manuscript for publication.

Every effort has been made to trace and acknowledge ownership of copyright. The publishers will be glad to make suitable arrangements with any copyright holders whom it has not been possible to contact. The author and publishers would like to thank the following for the use of photographs in this volume:

Lightworks Media / Alamy **p2**; Sparky/The Image Bank/Getty Images **p3**; Scott Camazine / Alamy **p6 (top)**; R D Battersby/BDI Images **p6 (bottom)**; AP Photo/NASA) /EMPICS **p7 (top)**; PASIEKA/SCIENCE PHOTO LIBRARY **p7 (bottom)**; © Lucidio Studio Inc./CORBIS **p13**; Michael Kelley/Stone/Getty Images **p29**; © Canon **p30**; © PETE CRONIN / Redferns **p47**; Leslie Garland Picture Library / Alamy **p80**; ImageState / Alamy **p90**; Farmhouse Productions /The Image Bank/Getty Images **p96**; Maximilian Weinzierl / Alamy **p113**; © Canon **p114**; David Stares / Alamy **p126**; Mojo magazine: © Mojo, Emap Metro Ltd **p147** bottom; Guardian: Guardian Newspapers Ltd 2006 **p147**.

Screenshots reprinted by permission from Microsoft Corporation.

Investigating multimedia products

What you will learn in this chapter

You will learn what multimedia means and how to use multimedia in your work.

There are sections on presentations, websites and how to incorporate sounds, movies and animations.

This chapter will take you through the multimedia products you need to have access to. It also has a number of questions you should ask yourself when you look at multimedia products.

Investigating multimedia products

Over recent years the use of ICT has enabled designers to develop fantastic, and not so fantastic, multimedia products.

This book will help you to design and produce your own multimedia products, but first you need to be able to discern what makes a good package.

What is multimedia?

We are bombarded by multimedia content, websites, TV shows, mobile telephone applications and even in-shop advertisements. But what is multimedia?

Multimedia is the use of more than one form of communication medium: sound and movement, text and images, or a mixture of all of them!

In the past, many methods of conveying information used only one form of communication – usually speech. It was not until the invention of the printing press,

by Johann Gutenberg, that the written word became commonly available to the public. This meant that people started to learn to read, communication systems developed rapidly, and so did school!

Nowadays we don't just have books to learn from. There is a massive industry supplying multimedia and interactive training materials and learning resources. There are even resources for you to work through to help you to understand this textbook.

> Although multimedia can mean mixing text and images, on their own printed materials are not normally considered to be multimedia.

Most of these materials involve some sort of text and image based file, usually converted from a paper document so that it can be viewed on-screen.

Some of these materials are an excellent blend of different media, some leave a lot to be desired!

Types of products

The world of education has been swamped with multimedia products. Some of them fall into certain categories:

- **Training materials**: These are materials that show the user how to do something or go through a sequence of 'lessons' designed to cover a particular operation. This could be an on-screen demonstration of how to add text to an image in Adobe® Photoshop®, or how to build a website in Macromedia® Dreamweaver®.
- **E-learning packages**: These usually cover a greater depth and breadth of knowledge. They could take a user from a beginner to an advanced level in a particular area, such as the use of a particular application. They may also take a user through a complete course of lessons, right up to taking an examination or submitting coursework, such as the *European Computer Driving Licence*, or even DiDA!

There are also various applications used to deliver these materials. They may be delivered to the computer through a **Virtual Learning Environment**, as standalone packages on CD or over the Internet.

The growth in these types of materials has driven some of the development of **ICT** facilities in schools. They often need quite a high-specification machine to use, even though the actual process they are teaching may not require such a high quality machine – an interactive video showing how to add up a column of figures in a spreadsheet will need much more processing power and a higher level of graphic capability than the spreadsheet itself!

It is only when working with video or sound that you really need the functionality and capability of most modern school machines.

Outside of school, there is a whole world of multimedia, perhaps the most interesting being in the field of entertainment.

Over the last decade the growth in electronic games has been phenomenal. Microsoft® predicted in 1999 that the major growth area for computing was going to be the games console and home entertainment area. In 2005, the three major games console manufacturers – Microsoft® (Xbox®), Sony® (PlayStation®) and Nintendo® (GameCube®) – all started to market their next generation consoles.

These latest systems will be able to deliver incredible quality images and interaction for a fraction of the cost of a similarly specified **PC**.

There are also hand-held devices, such as the PSP™ or Game Boy®, that offer mobile gaming on a level that matches much of the ability of a desktop machine of five years ago.

However, this does not mean that the games have improved. Many game reviews comment on the quality of the interface, or the quality of the imagery, but quite a number will also comment on the lack of originality or poor gameplay.

However, entertainment does not come only in the form of games. The avaliability of films on **DVD** has resulted in many shops not selling video tapes. Although tape can show multimedia applications, due to the linear way in which tape works – it has a start and an end – it has proven less popular. As the data on a DVD can be accessed at any point almost instantly, it has enabled developers to add material to enhance the main element: the film. This enhanced content has also led to an increase in the functionality of the DVD player.

Most DVD players can cope with sounds and images, but some can add quality to the sound, converting the signal to stereo, 5.1 or even 7.1 surround sound. They can also

interlace the picture (making the image appear better quality on the screen). Some DVDs contain games or puzzles that can utilise the software in the player to allow a user to play within the film environment.

Standard DVDs hold about 4.7 GB of data (approximately 4,700,000,000 bytes of data or approximately 37,600,000,000 1s or 0s!). The latest DVDs, such as Blu-ray discs, will hold many times this. The greater capacity will allow developers to add more content. Film companies are already experimenting by adding a range of different endings to films allowing an audience to influence how a film ends!

With the advent of interactive digital TV there will be a massive increase in the functionality of the television, as it will be possible to add content and interactivity to the images and sound.

Most UK homes now have access to digital television, either through satellite or cable. This has already changed the way many people *watch* television. Most TV channels have associated websites, and many of the programmes shown on the TV have their own websites or microsites.

The channels often deliver content which can be accessed through the digital system, in addition to the programmes. For example, in the past, sports channels broadcast one signal from a sports event. There may have been a number of cameras, but only one at a time would transmit to the viewer. Nowadays, there are still a range of cameras, but the viewer can select which camera they want to use to watch the activity.

Another aspect of television that has changed recently is the ability to take part in the programmes themselves. For example, audiences and viewers alike can take part in selecting the success or failure of competitors or participate in quizzes. This interaction began with the studio audience, but has now progressed to telephone voting, texting, using the interactive button on the remote control or visiting a particular website.

Interactive programmes are massively popular, and TV stations and telephone companies make a fortune from telephone voting.

There are also many types of multimedia products produced alongside TV and films, for example, games, audio books, etc. The ability to present pictures, text, sound and moving images has meant that there is no escape!

All organisations need to get their message across to potential customers. The development of multimedia products and the equipment to present these products has enabled this means of communication to grow enormously.

Many companies produce a catalogue or other promotional literature. Over recent years, some companies have started to develop these into websites or to present the information in a multimedia-based environment, such as a CD with images, text, a searchable catalogue or even showing short films of their goods being used.

Many schools and colleges produce annual prospectuses that are distributed in the local community or to potential learners usually; paper booklets. However many centres have moved to online prospectuses, accessed through their website. Some also distribute a CD which contains the same data as the paper version, but has additional information that cannot be shown on paper.

There are also many organisations that use multimedia presentations to convey information to others. Some devices have very complex sequences of events that need to be carried out in order to do a particular task. 'Help' systems can be developed to aid a person operating a device, for example installing a printer on a computer network used to involve reading pages from a manual, now it usually involves watching a short animation on a computer and following the on-screen instructions.

Automated Teller Machines (ATMs) sometimes have short films explaining how they are used. This helps the user and eliminate possible errors.

Multimedia presentations have meant that whole industries have sprung up to deliver the 'message' to the customer. It has also led to the opportunity to present materials for public use in a more accessible format.

Some people have difficulties with reading throughout their lives. There are also many people who are visually impaired. The development of the audiobook has been a fantastic way of allowing these people to access materials they would otherwise have difficulty 'reading'. Audiobooks are also great for drivers as they can be played through the car stereo.

E-books are another development enabling people to access books in new ways. An e-book is a copy of the text in a format be displayed on a screen, such as on a **Personal Digital Assistant** (PDA). The files are similar to **Portable Document Format** (PDFs) files, and are displayed through a specialist application, such as Microsoft® Reader. The reading software remembers where the reader finished and

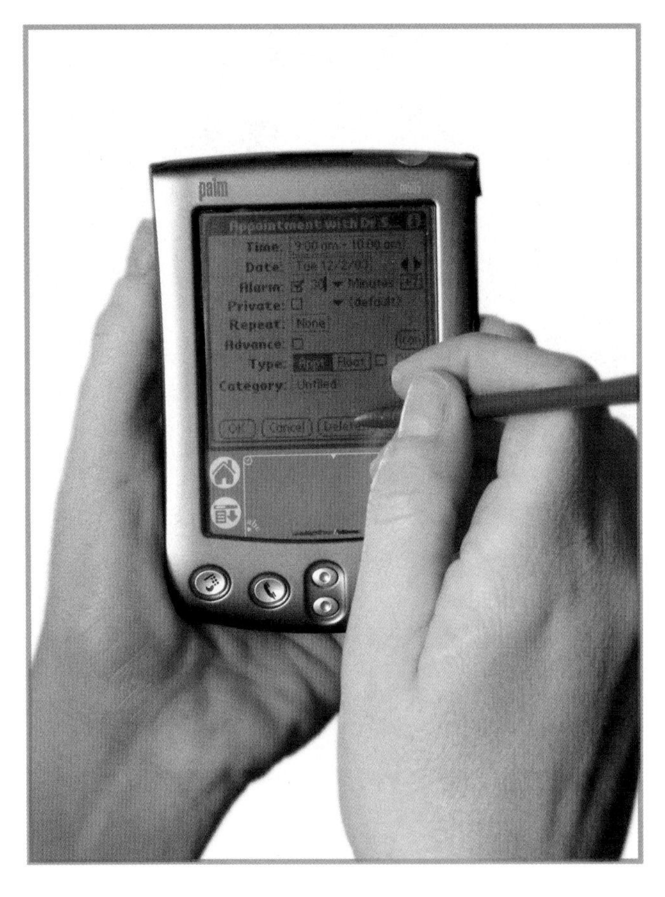

can also store any notes or marks that the reader adds to the text. Another advantage of an e-book is that reading software can often be made to read the text aloud. The only disadvantage is that the voice is not very good.

Due to the ability to store enormous amounts of data in ever smaller packages, it is easily possible to carry the entire contents of a public library in your pocket. An average e-book is approximately 500 kB, so it would be possible to store over 2000 books on one CD, or over 3000 on one 1 GB memory card. An Apple® iPod® could carry almost 500,000 books!

The ability to display the printed page on a screen does not mean that the file will suddenly become multimedia, but some publishers have managed to be creative and innovative with their material and have produced on-screen magazines (or e-zines). These contain text and picture content similar to a traditional paper-based magazine, with the addition of moving images and clickable content.

More and more people are also publishing their own material. There are millions of personal websites, however a website takes a certain amount of skill on the part of the designer. There has recently been a move to the development of low maintenance, easy to develop, sites: the web log (or blog).

Blogs are really online diaries. The person producing it can add text, pictures, short films or any other file, then publishes it on the Internet. This is usually done through dedicated blog sites, such as Yahoo!® or Microsoft® MSN®, where others can view their work.

Virtual systems

Virtual systems include virtual tours and simulators. A virtual tour is a video or sequence of stills taking the viewer along a particular path or through a set of rooms. Many hotel websites have virtual tours of their facilities. Also, many organisations have developed virtual tours of their premises for use as training material.

Some simulation devices use virtual systems to enable a user to carry out a sequence of operations without actually using the *real* equipment. Imagine learning to fly the space shuttle by sitting in it and being launched into space. What would happen if you couldn't do it? It is much safer to practise in a simulator, then if something goes wrong it can be reset and the learner can try again – with only their pride injured!

Many surgical procedures can also be simulated before being carried out on a patient. This has saved lives. There are occupations that use simulations because the activity is dangerous or may involve hazardous materials, such as working with nuclear materials.

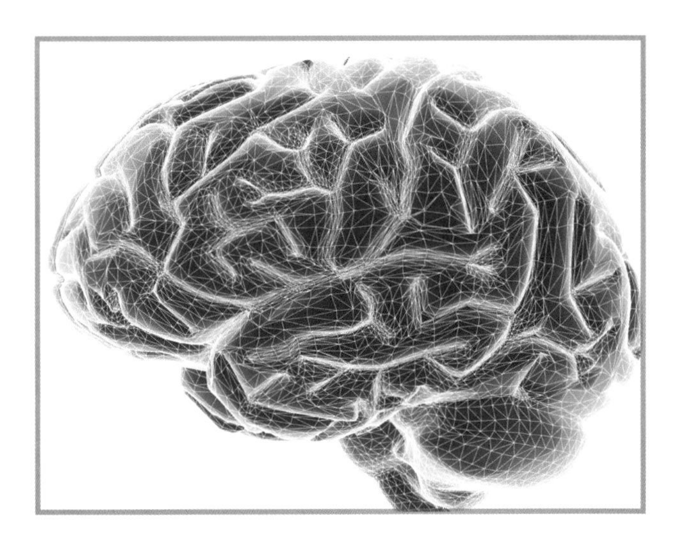

The processing power of modern computers has enabled three-dimensional modelling to become a very common way of prototyping items before manufacture. In the past models would be made from a variety of materials to show the designer what the finished item would look like, but computers can now be used to generate an on-screen three-dimensional image. The image can be rotated, scaled and coloured just as a model would have been. If a physical three-dimensional model is still needed the data can be sent to computer-controlled machines that can then make the model.

For many years software and hardware manufacturers have been trying to launch virtual systems for home users. The idea of wearing a headset and having parts of your body wired up to the processor has been around for a while, but no company has managed to perfect it yet. Instead websites have sprung up that offer users the opportunity to place an ***Avatar***, or three-dimensional representations of themselves, in a virtual environment. A very popular site is Habbo Hotel™. Here a user can move

from room to room or all around a hotel complex, meeting with others, chatting, and otherwise interacting, without leaving the security of their own home.

Public access

More and more organisations now wish to promote an image. Text can tell only part of the story, but the use of multimedia products has improved the accessibility of promotional material. Short film sequences or the use of animation and sound can help convey a message to the public in a more 'user-friendly' way.

There are a number of laws that now cover these sorts of materials. For example *The Disability Discrimination Act* requires that any publicly displayed material must do as much as possible to make the materials accessible to all, regardless of any disability. This means that, where possible, audio should be provided to help people with hearing difficulties, high contrast images and text should be used for those with sight problems and alternative methods of entering information or interacting with on-screen material should be offered for those that have problems using a standard keyboard and mouse.

Many organisations now actively encourage all members of the public, regardless of individual needs, to have access to their materials. This could mean having alternative material accessed from a common portal or just careful initial design.

Evaluating multimedia products

For your DiDA qualification, you must have an understanding of all these systems. That does not mean you need to be an expert in developing these materials, but you do need to know what makes a good presentation and what makes a bad one!

You need to be able to evaluate a range of material and suggest improvements. You will also be asked to design and make a number of presentations. You should apply the same evaluation criteria to your own work.

When investigating multimedia products you need to have a number of questions in mind. These questions form a framework for you to use for evaulations.

Access

- Can the materials be accessed easily?
- How does a user get access to the material? Do they need any security knowledge?
- If it is over the Internet, how long does it take to download?

- If it is via an on-screen display, is the display at a comfortable height? Can all the users see it without struggling?
- If a user has a disability, what are the alternatives?

Ease of navigation

- How does the user move around the presentation?
- Is the navigation method automatic or does the material wait for user input?
- Can a user navigate around the material without specialist knowledge or training?
- Are there helpful hints to support the user?

Content

- Is the content appropriate to the information?
- Is the content appropriate to the user? Age? Gender? Race?
- Is the language used appropriate to the audience?
- Does the user get the required information from the material?

Impact

- Does the presentation have an impact on the user? Will they remember it?
- Is the impact what was expected by the developer?

Interactivity

- Is there any interactivity?
- Why is there any interactivity?
- If there is, does it help deliver the information?
- Would more interactivity help? Would less be better?
- Does the user know how to use the interactive aspects? If they don't, how can they find out?
- Are there hints and tips?

Use of colour

- Is there a colour scheme?
- Does it match a 'corporate' scheme?

- Do the colours help convey the message?
- Are they necessary?
- Is there a version using other colours?
- Are colour-blind users catered for?

Variety of media

- Are a range of media used: text, graphics, sound, video, animation?
- Do the different media complement each other?
- Is the use of different media helpful?
- Is the use of different media necessary?

Techniques

- Does the material reflect skill in the designer's use of techniques?
- Are advanced techniques used to good effect?
- Are there techniques not used to the full?

Finally, you should consider the 'fitness for purpose'.

- Does the product do what it is supposed to do and does it do it in the most appropriate way?

Throughout this book, you will be looking at the work of others, studying professionally developed products that have been developed professionally and helping develop materials of your own or with your classmates.

At each stage, you should consider the questions above: particularly the last one. It is very easy to be convinced that something is useful because it looks good, but you need to have a more critical approach. Don't be tricked into thinking that because it looks good, it is good!

How to get good marks

✓ **You need to make sure that you record your investigations and use the most appropriate items from your searches.**

Homework

1 Collect three game reviews of computer or console games. Try to get screenshots of the games or scanned images, and produce a review of the graphics. Does your review match with the review of the game?

2 Produce a short presentation on the life of Johann Gutenberg or Thomas Caxton (printers).

3 Put together a short presentation on the advantages and disadvantages of digital interactive TV.

4 Produce a multimedia showcase of your skills, present it as a web-based presentation you could show to a potential employer.

Multimedia products

What you will learn in this chapter

You will learn how to produce multimedia projects. There are skills sections on incorporating clip art, animations, images and text for your target audience.

What is a multimedia product?

During your DiDA course you will work on multimedia projects. This means you will use software that allows you to incorporate and control sound and moving images. You can speak to, see and hear people on your computer as most modern computers have audio facilities (speakers and microphones), videos or DVD facilities. You can record sound and import movies from cameras, and download music or films from the Internet.

In essence the word 'multimedia' means a combination of sources: a mixture of sound, text and moving images which make the most of modern software features. Most modern PCs have speakers as a standard component of the package and these ensure that you can hear the audio tracks, whilst the affordability of digital cameras and video cameras has ensured that images (still and moving) can be captured with ease.

We ourselves communicate using multimedia means by using hand gestures or facial expressions when we tell a story to other people.

Modern PCs are capable of handling multimedia to differing degrees – sound and graphics cards can enhance the performance of a computer and ensure that a package runs efficiently.

The word multimedia has actually changed in meaning as it was originally used to describe distance-learning courses that made use of a range of sound and video-based components. As with most elements of the ICT world these days, the term has now changed beyond all recognition as it has now come to mean any way in which audio, video and text are combined to carry out a task.

A multimedia application would typically feature:

- The use of digital technology.
- Interaction with the user.
- Seamless integration of sound, text and data within the application.

Why use multimedia?

Multimedia packages provide the user with the opportunity to watch and listen to content that can help to make information more interesting than it might otherwise be. Such presentations can help motivate the viewer as the combination of visual images and audio can be more engaging than plain text alone.

What is a multimedia program?

The main components of a multimedia program are:

- **Text**: needs little explanation – words and numbers depicted in a range of font styles and weights (i.e bold, underlined, italics, normal).
- **Images**: for some people a picture speaks a thousand words – so a visual depiction of an event or object can be clearer than a textual description.
- **Moving image**: images of pre-recorded events can be shown. This can include one-off, unrepeatable or dangerous events.
- **Animation**: the advent of Macromedia Flash® has made animations commonplace on many websites.
- **Sound**: sound can be in one of a range of formats and can portray a whole range of emotions in ways that text or images could not.

- **User interface**: the user has to be able to navigate within the application. Often such features include 'skins' that add a sense of identity to the package.

These features can be combined within a single application:

- Clicking a static graphic in order to launch a more meaningful animated sequence.
- Clicking a block of text to launch a video clip.
- Clicking images of national flags to launch audio files in a range of languages.

When you are designing your project you must always consider your target audience. When you have identified your audience you will be able to produce output that will stimulate and interest them.

Time is of the essence, so you need to make sure that you use it wisely. It is easy to waste time on irrelevant aspects of the project, aspects which will gain you no marks.

Therefore you must make sure that:

- The application functions as required: your audience will not want to sit in front of a blank screen.
- Your user has to know what they are able to do – therefore all interactive buttons must function as you want them to.
- If the application is supplying feedback to the user, then it must be accurate – the user cannot be told they are incorrect when they have given a correct answer.
- The user must be able to navigate freely within the application.
- There should be an indication of the duration of specific sections so that the user knows they have time to begin a particular section.

Basic features of design

The project should have a logical structure so that the user is able to access all available information without difficulty. If the objectives are clearly identified then the user will be in no doubt about whether or not the objectives have been achieved.

The multimedia design model

The design model comprises four major functions:

1. **Analysis**: Think about what you need the application to do.
2. **Design**: Plan and lay out the different parts.

3 **Production**: Actually constructing the application.

4 **Evaluation**: Look at what you have done and check you have achieved all the goals you set out to do. If there are any parts that need modifying, evaluation will bring these to your attention.

Additional activities not included in this model may be required. The model is only meant to suggest the types of activities to be followed. It is a guide, not a blueprint.

Instructions

The multimedia design model should be adapted to describe the instructional design and development processes relevant to your project.

1 Conduct a 'Needs Assessment'. Who will view your multimedia work?

2 Write a 'Needs Assessment' report. Write down your findings.

3 Prepare an 'Audience Assessment'. Understanding your audience.

4 Include 'Audience Profile'. Age, gender, employments of your audience.

5 Specify 'Content and Objectives'. What you hope to achieve.

6 Write 'Content Outline' plan. What is in your work?

7 Select software. Which software will do the job best or which one do you know best?

8 Select 'Delivery Systems'. What will your work be shown on?

9 Write 'Instructional Objectives'. Are there any points you wish your audience to remember?

10 Plan the project. Plan what you have to do within the time you have.

11 Write 'Planning Evaluation Strategies'. How will you know if your work is successful?

Project timetable

What have I got to do?

- Screen design
- Flowcharts
- Scripts specification

- Screens formatting
- Improvement specifications
- Formative reviews
- Graphics
- Production
- Video/film
- Optical media
- Audio
- Video masters/CD-ROM
- Project documentation
- Evaluation report

Identifying your audience

You must start by identifying your target audience. You can establish this by asking a series of introductory questions, which can help you to ascertain their requirements. You should then write down these requirements as this will provide a good starting point for your project.

Ask yourself: What do my audience want to see in the presentation?

Now that you know what your audience wants, you need to decide how you are going to let them see and hear it. You need to bear in mind that sometimes there are constraints imposed on you such as bandwidth issues over the Internet. So although you may want to be ambitious, you need to make sure that it is realistic to attempt the task you are setting yourself.

Ask yourself: What is the best software with which to tackle this task? Will my audience be able to access the presentation finished? Do I have the skills that are required to use the package I want to use?

It can be very hard to maintain your audience's interest with long pieces of text on web pages. Sometimes it is suitable to present text in this way, for example in academic journal entries. However, for the most part you should stick to short manageable chunks that can be read easily – preferably on a single screen that does not require the user to scroll downwards or sideways.

Ask yourself:

- Can the information be logically presented in categories?
- Do the categories have a specific order i.e. Are some more important than others?
- Can text be unique? No one wants to read text that keeps being repeated.
- Can the categories be presented in a logical and structured way which makes sense to the user?

Organising the structure

It is important that your users are able to navigate around your presentation easily. If they cannot find their way around it, then they will give up! You can help yourself at this stage by:

- Clearly identifying all options to your users.
- Ensuring that the structure is interesting.
- Following a storyboard. This will ensure that the structure is visualised by you and you can follow this as you work on your presentation.

Presentation design

When constructing your presentation, you should:

- Identify a style for your work – this should be maintained throughout the presentation.
- Don't run before you can walk. Keep the task simple to begin with – you can always add layers of complexity later on.
- Create a scaled down working sample so you can see what the finished product will look like.

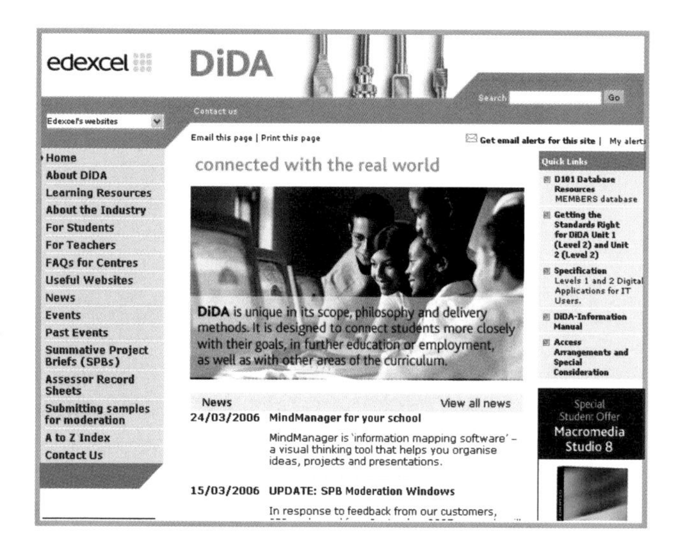

Visual style

It is important that you stamp your own identity on the presentation you make. This will help to make it stand out from other presentations and if you produce a series of presentations, the user will come to know what to expect from you.

This 'identity' can be achieved through the selective use of a range of techniques such as font styles and use of graphics.

Graphics

The first step is to choose a **house style** for your work. This gives a general theme to your work and will ensure that it is obvious to the user that all presentations in the series come from the same 'stable'. For example, you may have an orange border around each page, you may have a similar logo on each page or you may use the same, distinctive font on all of your pages.

It is a good idea to make sure that your colour scheme is complementary. There is nothing worse than presenting information in a colour scheme that a user dislikes. You need to ensure that you do not choose colour combinations that are hard to read e.g. red writing on a purple background.

All graphics files should be 'optimised' for their purpose, generally the smaller the better but not so small that all sense of definition is lost. The size of an image needs to be taken into account when it is being inputted into the computer – whether through a scanner or digital camera.

Text

The font you choose can make or break your presentation. You might have a great idea for using a swirly, calligraphy-type font – however if it is illegible then it might backfire, as your user may be unable to read what you have written.

Be sure to keep your sentences short and to-the-point. If you fail to do so it is possible that your user may lose concentration and at that point they will disengage from your presentation.

Media

Whether or not to incorporate sound or video within your presentation is one of the most important considerations you will have to make. You must ensure that they genuinely have a role to play, otherwise they may just be seen as gimmicky features added for the sake of it. Such features can also inhibit the performance of your presentation, so you need to be sure that you have carefully thought through every aspect of including such files.

Ask yourself: Is the presentation visually appealing? Can my user identify the content or does my style get in the way?

Layout

It is will help your user if each page has a similar structure as this will ease navigation – no one wants to struggle to find where to go next. If you are incorporating tables you should make sure that all tables are presented in the same way.

Limit yourself to as few font styles as possible – a total of no more than two or three is advisable except in exceptional circumstances. Beyond these few fonts, any others should be introduced only sparingly.

Bear in mind, when designing the layout, that some users will want to print the content. To this end, there needs to be sufficient contrast between text and background. A brightly coloured background will use up a lot of ink if printed in colour.

It is useful to show your user that more features are only a click away, so have links to other content clearly visible.

Ask yourself: Is the page logically laid out? Does the structure make sense to the user?

Interface design

Background: this is the most noticeable aspect of your presentation. As it fills the screen it will be the canvas on which the rest of your presentation will appear. It needs to be simple and not get in the way of the text and images that appear on top of it.

Balance: you should ensure that the layout of the page is properly balanced, rather than having all images at the top or always at the side. It is easy to engage your user simply by getting this aspect of your layout right.

Controls: make sure that all buttons follow a consistent theme it would be wrong to use a red arrow on one slide to denote moving onto the next page, and have a blue rectangle on the next page to denote the same action. Such inconsistencies will leave your user confused.

Video and sound: you need to make sure that these have controls attached to them. Users dislike video and sound files running automatically, particularly if they know their computer is unable to handle such files. For example a sound file playing automatically would be no use to someone whose computer has not got speakers attached to it.

Animating the screen: animated gif files can be used to add animated routines to your presentation. Otherwise you will require Macromedia flash and Shockwave skills to be able to create animations of your own.

Ask yourself: Will the user find the page attractive?

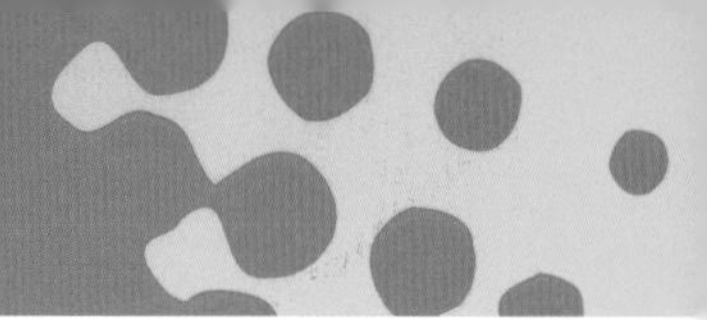

Readability

We need to understand that each and every word we present in a document is of great importance. It is up to us to ensure that your audience is able to make sense of what we have written. If we fail do so, we risk the possibility that the user will miss something vital and that could be catastrophic.

'Typography' is the term which describes the way in which text is displayed.
To use typography well we need to make sure that the text is legible and to do this the font we use has to be clear and unambiguous. It is also important to consider the colour of the font against the background when determining the readability of text.

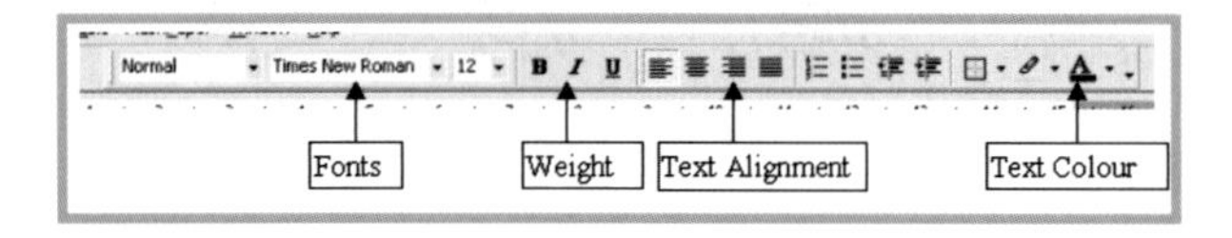

The following paragraph contains exactly the same text as the preceding paragraph, but is harder to read because the font is a scriptive typeface.

'Typography' is the term which describes the way in which text is displayed. To use typography well we need to make sure that the text is legible and to do this the font we use has to be clear and unambiguous. It is also important to consider the colour of font against the background when determining the readability of text.

If you can establish a good combination of text and background colours that work for you, you can hook your audience. You just have to make it as easy as possible for the user to read what you want them to read. The harder you make it for them to do so, the more likely they will be to stop reading your presentation. You will then have failed to meet your objective.

Although it is advisable to mix text with graphics wherever possible, it is sometimes not possible to do so. It is your task to grab the attention of your audience and if you only have text with which to do this, then you need to be creative. This is where the choice of font size and style can be crucial if you inadvertently choose a font style that is hard to read, your user will give up. Likewise, the wrong size can have unfortunate outcomes – too big a font size and the reader will think the text is too childish for them, too small and it may be daunting, taking too much effort to make reading it worthwhile.

Typefaces

Readability on screen

Most fonts can be divided into 2 categories:

- Serif
- Sans serif

Serif Sans serif

Serifs are the little 'flicks' that appear at the top and/or bottom of letters in some fonts. Serif fonts look good when shown in printed form but when rendered smaller on screen, the serifs can be hard to see – as a result webpages tend to use sans-serif fonts, with Verdana being the most popular.

Designed for the screen

Verdana is an example of a font that was specifically designed with the computer screen in mind. The structure of each letter has been optimised so that all characters can be easily distinguished, even at very small font sizes. In Verdana there is no confusion between similar characters such as '1', lower case 'L' and upper case 'L' or between 'i' and 'j'.

Choosing typefaces

Textbooks traditionally make a distinction in their use of typefaces. The main body text tends to be written using a serif typeface, such as Times New Roman, with category headings being written using a sans-serif font such as Arial or Verdana. This creates a distinction in the eyes of the reader and allows other forms of emphasis (bold, underline and italic) to be used within the body of the text instead of as headings.

Sound

There are many ways to produce sound for your presentation. You can record your own audio files by making use of microphones (inbuilt or external) with your PC or by using ready-made sound files. The latter are widely available on the Internet, but if you intend using audio files that you have not created yourself, you need to be fully aware of the copyright rules that affect such files.

Microsoft Sound Recorder®, which comes as standard on PCs is a simple but effective means of recording your own audio files. However, the software is limited in how it can be used to edit sound files. Far more effective for such editing is Audacity, which is an Open Source program that can be freely downloaded from the Internet.

Using Microsoft Sound Recorder

This package can be used to make simple recordings. A sound card is required, but as most PCs these days have one as standard, it is rare to find a PC that cannot make use of Microsoft Sound Recorder.

Once an audio file has been recorded, the software allows you to edit, mix and add your own effects.

Running Microsoft Sound Recorder

In Microsoft Windows XP®:
Go to Start > All Programs > Accessories > Entertainment > Sound Recorder

In Microsoft Windows '98®:
Go to Start > Programs > Accessories > Entertainment > Sound Recorder

The Microsoft Sound Recorder window resembles a cassette recording interface. A flat green line in the window represents the sound as it is played or recorded. To the left is the position indicator, which displays sound to hundredths of a second, and to the right is the length indicator which shows the total duration of the sound file. Below both of these is the slide bar which shows where the sound is playing relative to its overall length. The slider can be dragged to move directly to specific parts of the sound file. The buttons that lie under this slider are standard in form and represent Rewind, Fast Forward, Play, Stop and Record.

Playing a Sound

1. On the File menu click Open. A dialogue box will appear.
2. Navigate to the folder containing the sound file you wish to play. Select the file and click Open.
3. Click the play button to start playing.
4. Click the stop button to stop playing the sound.

Navigating within a sound file

How to get good marks

✓ You need to make sure that your product is fit for purpose, i.e it is well designed and suits the audience you are designing for, that it has clear text and images and that the sound animation and music is appropriate. For this area you need to plan carefully and make sure you have considered the use of graphics.

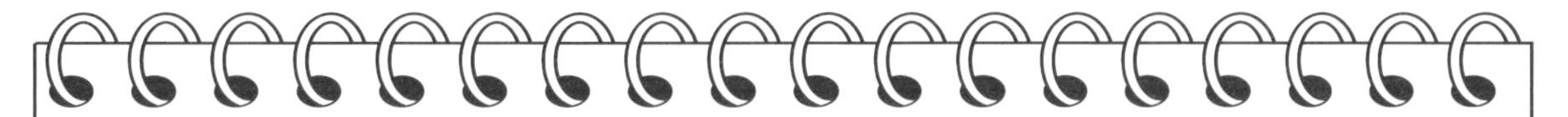

Homework

1. Using Microsoft® Sound Recorder and a microphone record a voice (speech or song), save it and play it back. Try to add a second layer of sound over the first using the tracks facility.
2. Write a few sentences using uppercase and then the same in lowercase. Try a sans serif typeface and a serif typeface on the screen and ask people to let you know which is easiest to read. Print out the same material and ask the same question.

Collecting and creating multimedia components

What you will learn in this chapter

You will learn how to find and use objects within your multimedia projects. There are specialist areas on working with images, bringing sound into your project and capturing images.

In developing multimedia products you will need to collect together a range of elements. These might be:

- Text
- Photographs
- Drawings
- Tabular Data
- Animation
- Movie clips
- Sound files

Some of these you can generate yourself. Others you will need to collect from elsewhere.

This chapter takes you through where to find things and what to do with them, and when you have found them, to make them ready for use.

Filing systems

In order that your materials are kept ready for use you will need to develop a filing structure. How to do this is dealt with elsewhere ((X-REF) to follow), but basically you should create a number of folders, ideally one for each sort of file you will be keeping.

Bear in mind that before you work with these files they may be relatively large, and your e-portfolio must be less than 15 MB, so you may need to arrange with your teacher or network manager to have a larger area for saving the files at this point.

You should also ensure that you have a backup of your files, both when you have edited them and before, so that if something goes wrong you can start again.

Images

Capturing an image using a scanner or digital camera is probably the best way to gather what you need. However you need to know how to use the equipment.

A scanner shines light onto an image and captures the reflection with an array of sensors. A digital camera works in much the same way.

A scan can be taken from within most applications. Microsoft® Word allows access to the scanning application: go to Insert > Picture> From scanner or camera.... Adobe® applications allow access via the File > Import.

Each scanner and digital camera manufacturer has developed their own software that is designed to work with a specific range of equipment. You will need to make sure that you match the correct software to the equipment.

Scanning software usually has a range of options that the user can set before scanning. This includes resolution, monochrome/colour and selecting an area of the image. When you have set these you can scan the image and then manipulate it in the 'host' application.

There is usually an option to save the image, remember to save an original in your images folder.

Digital photographs can be transferred to the computer in a variety of ways:

- Link the camera to the computer and download automatically.
- Remove the memory card from the camera and 'read' it with a card reader.
- Transfer using wireless technology, such as Bluetooth®.

Whether you scan or photograph an image, the trick is to collect relevant and useful images, and make sure you know how to use your scanning software and digital camera in advance of having to use it for your work.

If these methods are not able to generate the images you need, you may need to adapt an image or even draw your own.

Drawing applications offer similar tools, however some are much more flexible in what they can be used to generate. If you need a simple shape or cartoon image, it can be generated relatively easily in Microsoft® Paint, if you need more detail, use Adobe® Illustrator® or Macromedia Freehand®. Both offer professional standards of functionality.

When drawing directly onto a computer, you need to remember that computers cannot think for themselves (yet!). Before you do anything, make sure you have selected the correct tool, and chosen the colour you want to use for a line and for the fill of any shape.

You may have access to a 'drawing tablet'. This is a device that allows you to control the mouse with a stylus (a pencil-like device). This usually works just like a pencil, but as you move the stylus the cursor moves across the screen. The harder you press the darker the line drawn.

Simple animation

If you want to develop drawings into a simple animation, you will need access to an application such as Macromedia® Flash®. Although other applications can be used, Flash® is considered to be the most flexible and can be used to produce professional artwork relatively easily.

Animation is a great way of catching the attention of the audience, but don't overdo it!

The best forms of animation are simple. A shape 'morphing' into another, or a figure moving from one side of the screen to another.

To make a shape change or 'morph' you will need two states: a start and an end state. Flash® works with 'vector' drawings, so it may be best to develop your shapes in the applications mentioned above, although Flash® does have a full drawing toolbar.

1. Place the first shape on the Flash® canvas, where you want it to begin.
2. Extend the timeline to a time you think the change should finish and add a 'keyframe'.
3. Place the second shape on the canvas where you want it to finish.

4 Highlight the frames on the timeline and add a 'tween', either a 'motion tween', if the shape is moving or a 'shape tween' if the shape is changing.

5 Save your 'movie' and preview it by pressing F12.

Flash® is an extremely versatile application. If you wish to learn more, take a look at the Help files and tutorials that come with it.

There are a number of preset animations that you can use in applications such as Microsoft® PowerPoint®.

Experimentation is essential. Try a variety of methods to get the effect you want, but remember to record what you are doing.

Image type

Images come in a variety of types, some of which are easier to work with than others. You need to think about which format is best for your purpose.

If you have decided to generate your own images, you will have some control over their format.

You should always try to get images in JPEG format, although that may involve conversion from some other format, such as PSD or bitmap. Adobe® Photoshop® is able to open and convert just about any graphic format, so if you are having difficulties, try and open it in that application:

To convert a file to JPEG format using Adobe® Photoshop®

1 Open Photoshop®.

2 From the File menu choose Open.

3 Navigate to the folder containing the file.

4 Make sure that All Formats is selected in the Files of type: drop-down list box.

5 Double click on the file.

6 When it has opened, go to File > Save as and navigate to your photograph folder. Make sure the Formats drop-down box shows JPG. Give the file a name and click Save.

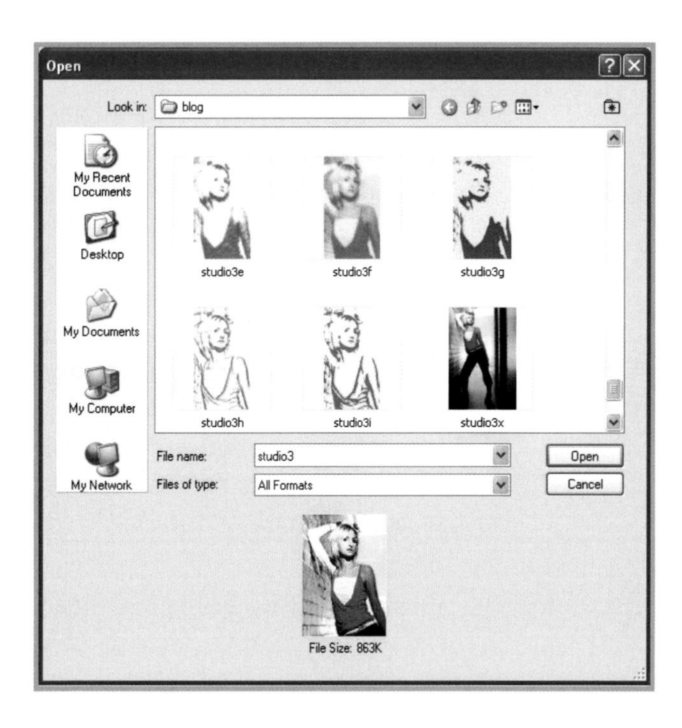

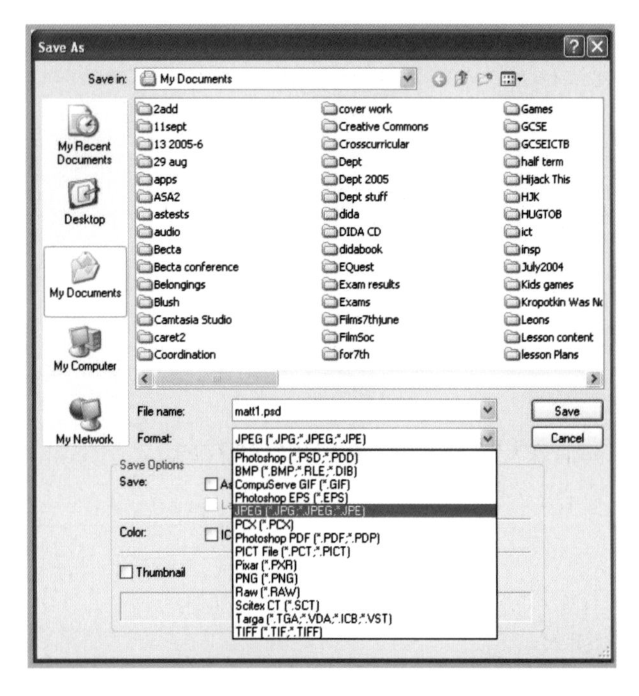

7 As the file is saved you will be asked to select a 'quality'. If the image is to be manipulated later, save it at the highest setting. If you do not intend to alter it later or it is to be sent over the Internet, you should select a lower quality, as it will keep the file size relatively small.

JPEG is the most common format for images. This means that most applications will allow you to import a JPEG file, without too much difficulty. Some of the other formats, including Adobe® PSD format, are not accepted by other software.

You will probably be gathering images from a variety of sources. This could also mean that they are a variety of file types. By carrying out the conversion above the images will have similar attributes and it will make your life a lot easier later!

Collecting images

Images are easy to collect. Unfortunately you may end up collecting images that you shouldn't use as some images have copyright. Before 'harvesting' images, make sure you are clear about whether or not you can use them. Look out for a copyright symbol on a website, or at the front of a book or magazine.

Some websites contain images that you are free to use. There are also thousands of free images contained on computer magazine cover discs.

You need to get the best quality image possible. If you come across an image you like, either on the Web or on a free image library disc, move the mouse cursor over the image. If you are using Microsoft® Internet Explorer or some other browser software you should see screen tip text containing some date or a title of the image. Right-click the image to get a small toolbar with options to save, print, email or open a pictures folder.

Click on the image, and if there is a better quality version available, it may open automatically. If your browser page changes, the image was a link, so click the Back button on the browser toolbar.

Right click on the image to bring up the shortcut menu. Choose Save picture as... . This will open a folder window. Navigate to the correct folder, give the image a name that you will remember and click Save.

The format of the image will be shown in the Save as type drop-down text box. To change this, click on a format on the drop-down list. You may not have any choice, in which case you will have to convert the file as described earlier.

Using clip art is another way of generating images ready to use in multimedia products.

The term clip art means a 'ready-to-use' graphic file. Clip art is usually installed as part of a software application on a computer, but it can be purchased or downloaded from the Internet.

Usually, clip art is artwork rather than a photograph. The illustration may have been drawn, converted from an original, or it may be an original.

The full version of Microsoft® Office has thousands of clip art images. They are accessed through the Insert menu. If you are unable to find an image that matches what you need, there is an online library of Microsoft® clip art that contains thousands more images.

This and similar libraries are searchable by entering a keyword. The system will return any image that uses that keyword in its associated data file.

Libraries are often copyright free and they can provide some great images for your work. However, to give your work a professional feel, you should avoid using clip art too much. You won't find much Microsoft® clip art on the IBM® web pages!

Video and audio

Video and audio clips are less often used mainly because the file size is so much bigger. It makes the files less easy to distribute.

As with clip art and other image libraries, there are specialist video and audio libraries. These can be found on the Internet and on cover discs.

Unfortunately another complication with these files is that they are rarely copyright free. Make sure you check before you use them! Millions of people share MP3 format files on the Internet, illegally. Each file being transferred is covered by copyright. Every

year there are hundreds of court cases around the world where individuals are sued for infringement of copyright. Don't be one of them!

Audio and video needs to be compressed for use in a multimedia presentation, so when you are looking for files, try to find already compressed versions. Look for:

Audio formats
WAV – approximately 10 MB per minute.
MP3 – approximately 1 MB per minute.

Video formats
MP4 – approximately 7 MB per minute.
Quicktime® (MOV) – takes about 6 times more space than MP4.
WMV – needs about half the space of MP4.
DivX – takes approximately the same as WMV.

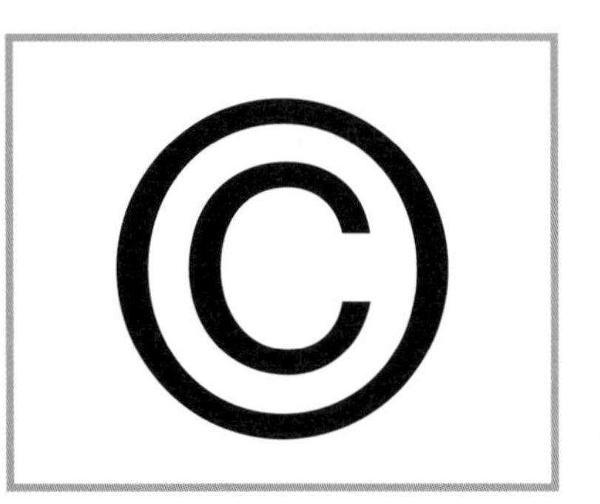

There are no hard and fast rules for using video and audio files. Some will import and work in some applications, others will prove difficult to import. This is because they need 'plug-ins' or 'codecs'. These are usually small bits of code that allow applications to convert the files for use.

If you are having difficulty importing a particular file type, take a look at the Help files. They usually list what file types are accepted by default and which types may need a 'plug-in'. The plug-ins and codecs are free downloads or may be available on cover discs from magazines.

Where to get video and audio

It is likely that any video or audio that you intend to use will need to be developed specifically for your needs. This means that you will need to record or film something and then transfer that to the computer. More detail on the intricacies of this is in chapter 12. However, there are all sorts of places that you can look for more general material or for material that you could not possibly generate for yourself.

If you need a short video of a news item, it is always worth taking a look at some news websites, for example:

www.BBC.co.uk
www.ITN.co.uk
www.Skynews.com
www.CNN.com

There are also a number of online archives or collections. Google® will search for these for you. You just need to enter a keyword or two. Google® also has a newly developed video archive of its own (video.google.com). It claims to be the biggest and is continuing to grow, so it's worth visiting if you need a clip.

Some websites offer free downloads of small videos suitable for showing on mobile phones and PDAs. These tend to be low resolution for viewing on with small screens, but often this is ideal for multimedia work, as it helps to keep the overall file size down.

There are also some small animations and very short video clips in the Microsoft® clip art collection.

Audio is a little easier to prepare for yourself. If you need a 'voice over' or commentary (someone explaining what appears on the screen), you can record your own voice using the software on the computer. The Apple® and Microsoft® Windows® have recording facilities. There are also very good applications available for download: Audiograbber is a sophisticated audio application that is free to download and use.

Bear in mind that sound can be very distracting. If you intend to use a 'voice over' make sure you follow some simple rules :

- **Script:** make sure you write what is needed – 'ad-libbing' is great if you are a professional, most people need a script.
- **Rehearse:** make sure you check that the 'voice over' matches the timings of the material.
- **A good voice:** although this is your work, it may be useful to have someone else to do the 'voice over'.
- **Recording:** make sure you work in a silent room – noises in the background will be impossible to remove later, and will be disturbing to the audience.
- **Tracks:** if you want to use music as well as voice, you need to use two or more recording tracks – this becomes much more complicated and you need to record the voice on one track and the music as a separate track, then add them together (Flash® and Director do this).

If you need audio material that you cannot record yourself, there are various websites that can supply MP3s of music, sound effects and other tracks. There are also discs of materials available from magazines.

Specific noises can be useful to help explain something or to emphasise an effect. For example, the sound of typing can be used when text appears on the screen, a gunshot can be used at the start of something, or a machine noise can be used to give the feeling of industry. These types of sound can be found in clip art libraries or 'ripped' from games and other multimedia material.

Whatever you choose to use, make sure you have permission.

Text

Perhaps the largest part of a multimedia product is text. This may not appear in the presentation, but it can be used to supply the script or give content and direction to the designer or developer.

Text can be generated through research and knowledge or can be 'harvested' like the other elements.

Unless you intend to develop materials that are very personal to you, you should search widely for appropriate text.

Again copyright issues may exist, so make sure you look for copyright symbols or other statements. But **generally** if you only intend to use the material for educational purposes, like your DiDA work, you will be working within the law. If you present your work for commercial gain (to make money), then you may be breaking the law, so make sure you check.

If you need text on a specific topic, use the Internet. Open a search engine and enter a few keywords. If you are producing a sales report, use words like *sales statistics* or *sales figures,* not just *sales*.

When you find some appropriate text, it can usually be copied and pasted into a document in the same way as any other text. However, web-based text has often been formatted in ways that do not work on a standard word-processed page. Text is often aligned in tables or has 'hidden' formatting that forces a browser to show it in a certain way.

To remove the formatting and make the text appear as you want it to, you will need to extract it from any tables and then highlight the text and apply a style that is more appropriate. This may cause oddities to appear: bullet points in web-based text are often small images or special symbols and not formatting. You will need to delete the bullets on each line, individually.

You can also scan in text from a magazine article or book. Although scanning software is usually used to get images into electronic format, most scanners also have Optical Character Recognition (OCR) software.

OCR software enables a scanner to scan the page and then convert the image of the text on the page into editable text that can be imported into a word processor.

Nuance TextBridge® is an application often used to convert scanned text for use in a word processor.

Start by finding a suitable page of text. Place it on the scanner and click **Get Pages.** The application will automatically scan your page and generate a **thumbnail** and a preview.

1 Highlight the text you want to grab using the **Mark Text** button, then click on **Recognize**. The highlighted text will then be converted and any words that the application is unsure about will be highlighted and appear at the top of the page. Go through these and accept or change them.

2 When you are happy that you have the text you need, click **Save As,** navigate to your text folder, check that the **Save as type** drop-down text box matches your requirements and give the file a name. Then click **Save**.

The application will then carry out the conversion and save the file. It can also open the file in your favourite word processor.

You will probably have to adjust the formatting.

Reformatting

Each of the items mentioned so far, apart from sound, will appear on the screen. This means that you will probably need to alter their appearance.

Images

The specifics of how to use particular applications can be found elsewhere, but in general, you will need to use the following operations:

- **Cropping:** Removing parts of an image by trimming the 'canvas' it appears on. This removes data, so it can be irreversible: Take care! It is useful for getting rid of material that is not essential or for 'tightening' an image.

Uncropped

Cropped

- **Resizing:** This alters the size of the image, but does not add or remove any data. This can make images 'pixelate' as increasing the size is the same as zooming in. You also need to be careful when resizing that you constrain the proportions otherwise some very odd effects can happen! Making the image smaller in Microsoft® Word will not make the file size smaller. To alter the file size, as well as the image size, you will need to adjust the size in an image manipulation application.

Original Larger (and cropped) Smaller

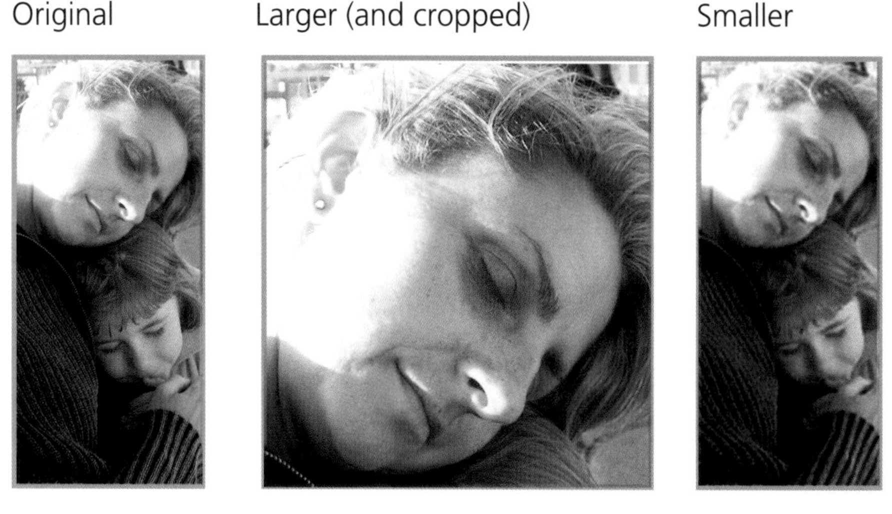

- Adjusting brightness and contrast. Although image manipulation can be carried out using most applications, the more specialised the application, the more variety of adjustments there will be available for use. Microsoft® Word allows images to be adjusted, but Adobe® Photoshop® has a far greater selection of tools and effects.

More Contrast Less Contrast More Brightness Less Brightness

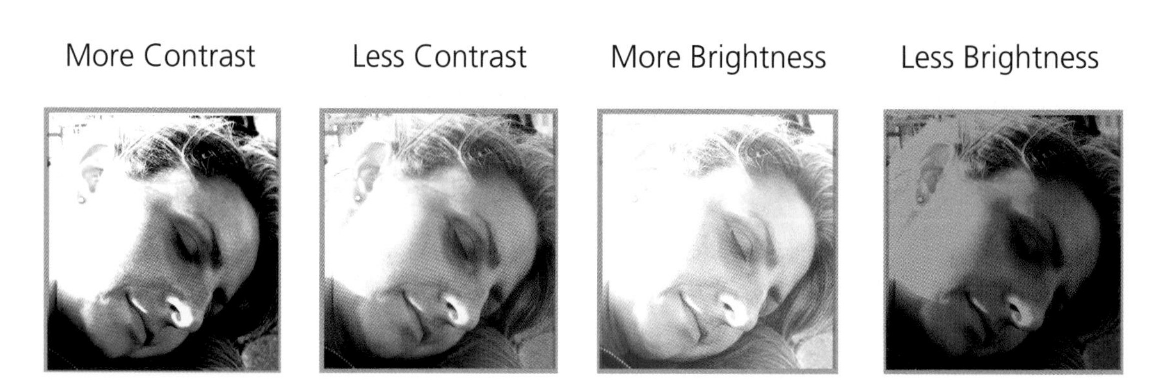

Altering colour requires a specialist application (software such as Microsoft® Word has limited functionality).

Within an application such as Adobe® Photoshop®, there are functions such as filters, which can alter the image in many different ways.

Think carefully about what you want to do with your image and make sure you keep the original.

Text

Text can also be resized and adjusted. Typically this requires the text to be highlighted and then formatted using commands on the toolbars. However, adding text to an image should be done in a specialist package, such as Adobe® Photoshop® or Illustrator®, as the options are much greater.

How to use a word processor is detailed elsewhere, but remember that when using different text formats, you should always try to use 'styles'. This helps to keep the file size smaller and it also makes it easier to make changes to documents.

As text will often be manipulated within a word-processing application, there are a number of rules to remember:

- Always use font colours that contrast with the background (except where a particular effect is required).
- Only use the Enter key at the end of a paragraph, not at the end of each line, as it is difficult to adjust the line length or font size if each line is a separate section of text.
- Get all of your text onto the page first and then do the formatting. It is much quicker.
- Always use the spell checker facility.
- Headings and titles should be clear and effective. Sometimes adding a drop shadow or some other text effect can help.

Video

Video is much more complex than images or text, although the basic effects of copying and pasting or resizing and cropping are usually the same as when working with images.

To carry out any further adjustments you will need to use a specialist application, such as Microsoft® Movie Maker, Adobe® Premiere®, Apple® iMovie® or another of these packages.

One of the most important things to consider when working with video is the size and resolution. A movie digitally shown at a cinema is projected onto a screen over 20 metres wide. The resolution of the image needs to be very high to give a crisp and clear image. A standard feature film for use at home can easily fit on a single DVD (less than 5 GB), a movie to be digitally shown in a cinema needs to be at a much higher resolution, with a consequent increase in file size. The same film will need more than eight DVDs, in excess of 40 GB!

At the other end of the scale, different applications use different means to compress video to different sizes. To view a short movie on a computer, WMA or DivX are the best formats for keeping the file size small. Both of these can be generated by the packages mentioned above.

When designing your multimedia product you should think about why you are using video. If it is essential, you then need to decide on how large the video needs to be.

Introduction
30 seconds

Presentation
60 seconds

Show video
45 seconds

Question
and answer
60 seconds

Try to avoid full screen video, running a movie in a small window might be all that is needed. By reducing the screen size of the video you can cut down the file size. However, adjusting the screen size of the video needs to be done in a suitable application, changing it in Microsoft® PowerPoint®, for example, will not reduce the file size.

Timing

Bear in mind that when working with multimedia applications, the greatest difference between these and single medium material, is the aspect of time.

Multimedia can be delivered sequentially that is, action A is followed by action B. This means that your planning must include a timeline. Each element needs to be given a time allocation and added to the timeline.

If a presentation needs to fill a specific period of time, you will need to build this in your timeline. Some of your elements may already have their own timeline (videos and sounds take time). These are factors that you need to consider.

Both video and sound can be manipulated to alter their timelines. Much the same as altering images, you need to be careful that you do not ruin the source material. By making a sound take longer to run, it will alter the sound itself. Making it run over a shorter time will make it run faster, therefore also altering the sound.

Very small alterations may not be significant, but if you need to make large changes, adding or removing a few seconds, you will need to edit the source material, not merely speed it up or slow it down.

Acknowledging the sources

The concept of copyright has already been mentioned. But, as you are studying for DiDA, you are allowed to use some material that would otherwise fall under copyright legislation. You can copy sections of text and images, use short videos and some sound files, but you still need to make sure that you get permission and acknowledge your sources.

This acknowledgement can be done in many ways. The easiest is to add a 'credits' page, just like those that appear at the end of films and TV programmes. However there may be other ways that are more relevant and appropriate, such as:

- add a copyright logo to the image or video;
- add a title to the image or logo;
- acknowledge the source in the text.

You will need to record where the material came from, who produced it originally and when it was produced.

How to get good marks

✓ **You need to make sure you focus on collecting suitable content. This could include, images, video or audio recordings. Try to show the examiner that you have considered things like colour and size, and make sure that you acknowledge your sources. Save your work in an appropriate format.**

Homework

1 Set up your filing system, to separate out your different files.

2 Collect at least one of each of these file types
- DivX
- MP4
- MP3
- WMA
- JPG
- BMP
- MOV

3 For each of the file types:
- record what application it can be opened with;
- record its size;
- make a note of any other important issues.

4 Produce a short animation of your name or a title for a presentation.

Developing multimedia products

> **What you will learn in this chapter**
> You will learn what frames and tables are used for in web page design. You will also learn about fonts, colour schemes, backgrounds, hyperlinks and how to link pages in your web pages.

Getting multimedia ready

Computers have to be able to cope with the rigours of processing multimedia files. The designer can aid this process by ensuring that all files are as compact as possible. The dilemma lies in ensuring the files are small enough to be manageable whilst being of sufficient resolution to be recognisable. This is called 'optimisation'.

You can tell when files have not been efficiently optimised by a designer if you visit websites where video files take a long time to load up. It is likely that such websites will not hang onto their visitors if the content takes a long time to load. You need to give careful consideration to accessibility and this means taking account of file types and formats. Sometimes you will have to settle for a solution which may fall short of your preferred outcome, purely because it works. One solution to help in this regard is to use frames.

What are frames and tables?

Newspapers are distinctive in that the text they contain is usually presented in columns. A newspaper layout makes use of columns and rows which are blocked to accommodate advertisements and photographs as well as text. This means the layout of a page is carefully considered and not haphazardly constructed.

You can think of frames as performing the same function as columns and rows. Frames can be used to divide web page content into separate areas depending on their function.

You can use frames to create web page layout in Macromedia Dreamweaver®, Microsoft Powerpoint® and Microsoft FrontPage®, although deciding which one to use can itself present a challenge.

A common use of frames is to separate the navigation menu from the rest of the web page content. It is useful to have the navigation within a separate frame because the user will be able to see it clearly without having to search around for the tools they need to move within the site.

Whilst frames and tables can each fulfil the same sort of role, there may be times when one is a better option than the other. For example, if the navigation options are to be placed at the left hand side of the screen, then a page can be split into two frames or a table can be created in the same position on each page. The crucial difference is that a frame is created once, whereas tables need to be made on each page.

Most pages are constructed from three frames:

- One thin frame at the side with a menu.
- One frame across the top that will contain the logo and title of the web site.
- One large frame that takes up most of the screen that will contain the main content.

Each frame is actually a separate page of HTML but with the use of framesets they collectively compose the entire page. A frameset is an HTML page that defines the main properties for the webpage:

- the number of frames on the page
- the source of the page which is loaded into a frame
- other definable properties.

The frameset page is not seen by the user, it merely contains the information required to build the page that the user does see.

The example page makes use of nested frames – this where a second frameset is contained within the first frameset, which is in turn held within the overall frameset.

Unfortunately, such structures are not without their problems as the designer has to keep track of all HTML pages that have been built, in addition to all of the frameset pages.

Saving frame-based websites

Users need to be in the habit of saving pages in a methodical way. If you can get into a routine then you will save yourself a great deal of trouble later on. Every HTML page and every frameset page has to be saved independently or the overall page will not work as planned.

How to save a frameset file:

1 Select the frameset in the Frames panel or the Document window.

2 Choose one of the following options:

a Save the frameset file by choosing File > Save Frameset

b Save the frameset file as a new file by choosing File > Save Frameset As…

To save a document that is inside a frame:

Click in the frame to select it and then choose File > Save Frame or File > Save Frame As…

To save all files in a frameset:

Choose File > Save all Frames.

This saves all documents that are currently open.

Note: Use the frame selection lines in the Doument window to help you identify frameset and frame documents during this process.

Formatting frames

The format of a frame sets the font type, font size, and alignment that is applied to the body of text within the frame. Each frame is formatted independently by using the Property Inspector which allows you to set such features as:

- Scrolling
- Width
- Height
- Page names

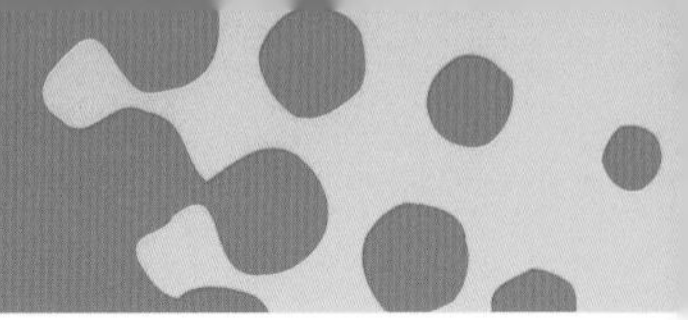

At this stage it is important that you name each page and it is helpful if you have a methodical system when you do so.

Frame design guidelines: Overview

Does your site need frames? These days people tend to use frames for gimmicky reasons rather than reasons of functionality. You should never do so when you make a page. On the contrary, you should give careful consideration to the needs of your site and only if you firmly believe that there is a need for frames to be used should you ever make use of them.

Thus when you are at this stage of the design process, you need to be sure that your user will benefit from seeing multiple documents at the same time. If you cannot establish the need for such provision, then you have no need to use frames.

Designing an introduction page

It is important to establish if there is a need for frames on the first page. You need not be aware that some users become irked by the use of frames and as soon as they see a frameset being constructed on the screen, they may navigate away from your site and go elsewhere instead.

One reason for this is that as frames are actually several documents, it can take longer for a set of frames to load than it would for an individual document to load. Some users dislike this and you need to be aware of this at the outset.

With the 'table of contents' frames application it can happen that the user can only navigate using the 'back' button on their browser instead of navigational buttons within the site itself. This is a sign of bad and inconsiderate design and could confuse a novice user.

Handling external links

It is vitally important that the framesets do not function when your user follows any external links. If they do keep functioning then the user will be unable to navigate out of your site and will be effectively stuck in an infinite loop. One other aspect of such a technique is that it can appear that the external content is part of your site and can lead to accusations of plagiarism.

Hyperlinks

Hyperlinks are objects that you can click on to transport you to a different web page or file. If you use these creatively they can be used to navigate around a vast array of content. Hyperlinks can be attached to a whole range of items, including:

- Text
- Images
- Shapes
- WordArt

As you start to make use of hyperlinks it is possible to lose track of where you are. You need to retain control of the structure of your site. In order to achieve this it is wise to create a storyboard in order that you can establish the destination of each hyperlink.

1 To make a hyperlink go to an object or piece of text and highlight it. Then go to Insert > Hyperlink ... The Insert Hyperlink dialogue box will then appear.

2 Locate or type in the file, or page or website you wish to link to and click OK. When you show a Microsoft® PowerPoint® slide the cursor will appear as a hand. If you click the link it will take you to the destination you chose.

To return to original text you were viewing, you just click on the Back button in your browser.

It is mainly due to the use of hyperlinks that the Web functions so efficiently. If it was not such an easy process to launch applications or visit other sites simply by clicking on a link, then it would not be so user-friendly.

More and more computer programs, such as Microsoft® Word, are making it easier for users to publish their own hypertext documents.

In summary, hyperlinks cause the selection of an object to result in a move to a new location or the performance of an action. This action can be any of the following:

- Navigation to a different slide in the current presentation, to a different presentation altogether, to another application altogether, or even to an Internet Web page.
- Run a program.
- Run a macro (a program written in Microsoft® Visual Basic® for Applications).
- Play a sound.

Adding sound from a file

1 If you wish to use a new sound in Microsoft® PowerPoint®, download and save the file on your computer, preferably in the same folder as your presentation.

2 Click on Insert menu > Movies and Sounds > Sound from File... .

3 In the Insert Sound dialog box, select the drive and folder where the sound file is.

4 In the file list, click the sound file you need, and then click OK .

5 Decide whether you want the sound to play automatically or on mouse click. If you choose mouse click, you will obviously have to click the icon with your mouse during your presentation to start it.

You can turn any object or text on a slide into a hyperlink as follows:

Right click on the object and select Action Settings... . (or select the object and choose Slide Show > Action Settings...).

In the Action Settings dialog box, select either the Mouse Click or Mouse Over tab. Select the radio button Hyperlink to and then choose the destination or action. (To link to the Internet, choose Hyperlink to URL and enter the web address starting with 'http'.

Compressing files

Compressing or making files smaller and easier to handle is commonplace when working with multimedia products. Files can become very large so compressing makes them quicker to send by email and saves space on a hard disk and network. The person at the other end can decompress the files on their own PC rather than having to wait ages for the files to download. The most common form of compression is to Zip®.

What are compressed files?

Compressed files are essentially folders in which the contents have been 'squashed' to save memory. This is particularly useful if you have a large set of files that you need to send electronically. A compressed file is like a vacuum pack of coffee from which all of the air has been removed to make the package more compact.

In order to make use of a compressed file, the sender and recipient both have to have access to the compression utility, the most common being:

- WinZip for PCs
- Stuffit for Macs

Why do people use compression files?

Using compression files, it is possible to send large files more quickly by electronic means than might otherwise be the case.

Typical uses are:

- Distributing files on the Internet: multiple files can be downloaded from a location in a single action.
- Sending a group of related files: when sending a Zip file it is as viable to send multiple files within a folder as it is to send single large files.
- Saving disk space: as a means of archiving files once they have served their purpose, Zip files can take up less room on a hard disc than in an uncompressed state. They can then be uncompressed if they have to be accessed at a later date.

Where does WinZip fit it?

You need to have access to a compression utility, such as WinZip, in order to make Zip files. Users of Microsoft Windows® can work with ease when using the likes of WinZip as the interface is straightforward and intuitive.

How do I open a Zip file?

The WinZip package is available initially on a free, 30-day trial basis. Once WinZip has been installed you can open a Zip file by double clicking the file whenever you see it – whether it is as an email attachment or as a file on a webpage.

Alternatively, you can double click the WinZip utility and choose Unzip or Install from an existing Zip file in the WinZip wizard. The wizard will then guide you through the rest of the process.

How do I create a Zip file?

To create a new Zip file you need to open WinZip. This package has two modes and for novice users the wizard mode is the most helpful. However, if you are using the trial version you need to be aware that the buttons for 'Use evaluation version' and 'Buy WinZip' switch positions each time you use the package – it is not just your imagination!

You will be asked 'What do you want to do?' You need to select Create a new Zip file and click Next. The WinZip wizard will guide you through the process.

Colour schemes

You should not underemphasise the role of colour when making content attractive to the user. If you use the right colour you can engage your audience and ensure that they want to read more of what you have got to say. Get it wrong and you might convey the image that your site is not interesting, and your user will go elsewhere.

It is sensible to spend some time planning a colour scheme that will add coherence and character to your multimedia project.

Viewing habits

Web-based applications are more flexible than paper-based documents in the way that data is presented. There is no absolute limit to the length of a page and there are no rules governing such limits either. It is up to you as a designer to present your information in such a way that your audience finds it interesting, engaging and attractive.

There are some principles that you can follow which help to attract your audience's attention:

- The most important content should be on the top third of the page. If it is not possible to do this, at the very least the content must not be located off screen as the user may not notice it if they do not scroll downwards or sideways.
- All links should be obvious. The mouse changes into a hand when it is hovered over a hyperlink, but using specific colours for hyperlinks can aid your user.
- Keep the most common options together. The features that you want your user to make use of most often should be easily identifiable and in the same place on each page. If the user sees a button in a certain place on one page, they will expect to see it in the same position on another page.
- Write for the web. You will find that the user who accesses your content via the Internet will expect the information to be presented in a snappier format than it would be in a book. Do not make it difficult for them to read what you have written.
- Use space wisely. It is not necessary to fill all space with text. The more spread out that text is, the easier it will be to read the information.
- Avoid busy or dark backgrounds. Your primary objective is to ensure that your user can read your content. If you put dark backgrounds or complex images behind the text then you will make it difficult for the user to do this.
- Remember the basics! It is easy to become engrossed in making use of the fancy features of the software whilst forgetting about the basic elements that your site needs. Remember to include such information as contact details so that your user can get in touch with you if necessary.

Colour schemes

We are all influenced by colour. Colour can give rise to certain emotions or be seen as fashionable, cool, hot, depressing, relaxing, etc. Colour can attract the eye to different parts of your information. It can make things stand out and become clearer. Colour can also define the 'image' you want to present. Many companies use colours to make their identity seem uniform.

Music

If you decide to add music to a web page or presentation make sure it is appropriate! Find a site that plays background music when you would not expect it. If you use music, make sure you offer the user the option of turning it off easily and quickly.

Pop ups

Pop ups are the topic of many a heated discussion about the Web. They annoy some of us with 'in your face' advertising. Many people block them appearing. Microsoft® Windows® XP and other security software tells you every time one is blocked.

Pop ups are big business and there are millions of them popping up every second on web pages all over the world. They are a bit like the Web's equivalent of junk mail that gets posted through our letterboxes each day.

How to get good marks

✓ **You need to show the examiners that you can use hyperlinks and that you have considered file sizes and download times and your product is fit for its intended purpose.**

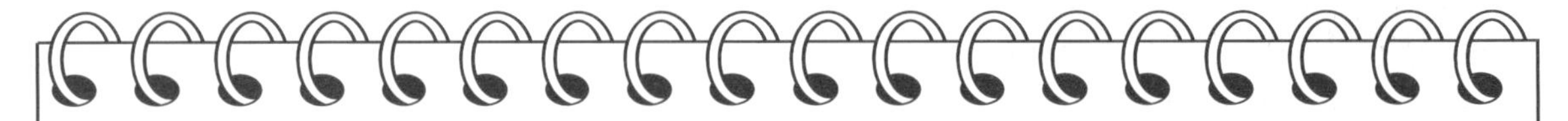

Homework

1 Try a simple hyperlink exercise in Microsoft® Word. Make a block, then go to the top of the screen and click Insert > Hyperlink.... Then make it go to a website you know, for example www.bbc.co.uk.

   Now try simple typing in a web site address, and click on it while holding down the CTRL key. What happens?

2 Load Microsoft® Word's web design tools in View > Tool box > Web tools. Create a simple web page then go to File > Web Page Preview.

Prototyping and testing

What you will learn in this chapter

You will learn the purpose of testing and modelling your product before you produce your final outcome. You will learn the importance of feedback and evaluating your work.

This chapter will try to explain why it is important to test your material and it describes some of the things you need to do before your work 'goes public'.

Presenting your material to an audience without testing can lead to disaster. You may have seen errors in books, magazines and even films. This can reflect a lack of care or an unprofessional approach. Make sure your work always looks sharp.

Fitness for purpose

If it is good at what it does, it will usually look good doing it!

This is the basis of the 'form follows function' concept. If you have done everything you can to the best of your ability, and your product meets the needs of the client, it will probably look good and work well.

If, on the other hand, you have cut corners, done the bare minimum and not quite met the design specification, you will probably find that your work doesn 't look very good, and fails to communicate what you intended.

Before you start any project, you should make sure that you are aware of what you need to do and that you are up to the task. If at the outset, you are not clear or feel that it is beyond your ability, discuss it with others, particularly your teacher.

It may be that you will need some help along the way or you need to learn some new application or range of processes, but this should not stop you.

Whatever you need, make sure you can get it, then approach the project with confidence and aim to do your best.

At the beginning of any activity you need to know what you are expected to produce. It may be that the final outcome is very much up to you, in which case the 'fitness for purpose' aspect is yours to decide.

If you are asked to present material to a particular audience you need to make sure you know as much as possible about the audience and what their tastes might be, as well as their level of knowledge and other points. You can then try to make your material match their needs and it will be closer to being 'fit for purpose'.

By now you should be familiar with the applications you will be using for developing multimedia products. But how can you get the best from them?

Quality log

Elsewhere in this book and in the other books in this series you will have come across the idea of using a Quality Log. This is a very useful tool.

You need to record your early attempts at satisfying your clients' needs. The following pages take you through some of the testing and modelling that you should do at an early stage in your project.

You should record your successes and your failures. Your Quality Log can be used to record when you achieve your objectives, as well as recording potential problems and possible solutions.

Prototyping

At the end of this prototyping stage, you should be able to put together a plan that you are sure will be successful. Your Quality Log can be an essential component of that plan.

Prototyping or modelling is an essential stage in developing multimedia products.

It is important that you check the ideas you have can be put into production. This stage can save a great deal of time and wasted energy later.

Imagine building a jet plane without having tested it, or even a new pencil case without making sure people would like it!

A prototype should test the important aspects of the product. If you intend to present your product in a certain way, such as on the Internet, you need to test that it can be accessed in that way.

If you intend to have a presentation that lasts a particular length of time, you need to test it, and adjust it so that it fits to your time slot.

Although you will test most of the prototype yourself, it would also help to have a range of other people to offer comments. Try to get a range of abilities.

What to prototype?

At an early stage your prototype needs to check a number of things:

- Timing
- Content
- Layout
- Colour scheme
- Transferability
- Interactivity

Later there will be other aspects to check.

Timing

You need to make sure that each of your elements will fit to your timeline:

- If you are using an audio backing track, you need to ensure that is it long enough or can be 'looped' so that it continues to play for the duration of the presentation. Bear in mind that looped tracks, particularly short ones, can be very annoying after a while!
- If you have a video sequence make sure it is long enough to show what you want to show, but not so long that the audience lose interest. A short, appropriate sequence is far better than a rambling movie!
- If you have actions taking place at particular points you need to ensure that they are 'triggered'. At the end of a video, what will take its place? The audience will not be very inspired if they are left looking at a blank screen!

Content

Although the product is at an early stage, you should be able to test some of the content:

- Test your images. Do they look good on display?
- Use Lorem Ipsum text (a dummy text that looks like real Latin words) to check for text flow around objects.
- If you are using numerical data, model some different graphs or charts and show them to other people to see if they can understand them.

Ask yourself: Is the content interesting? If it doesn't interest you, how will it interest the audience?

You need to check that the information you are trying to convey is interesting and engaging. Your job is not easy! Companies spend millions of pounds on advertising, much of which is specifically designed to be interesting and engaging. Now you have to do the same.

There are a few tricks to keeping peoples attention:

- **Keep things moving:** Don't leave the screen unchanging for more than a few seconds (unless there is some audio that accompanies the static image).
- **Don't move things too quickly:** If there is something that you want the audience to read or watch, give them time to do so. This may seem to contradict the rule above, but you need to strike a balance.
- **Don't have things happening in two places at the same time:** Have only one movie running at a time or only one animation happening. The only exception is sound, where it is sometimes a good idea to have a background sound as well as a voice over (although this can be distracting).
- **Don't frighten the audience:** If something suddenly appears on the screen the viewer will take a couple of seconds to register it. If it is a video that means they will miss the first few seconds and they will be confused for the rest of the show. Make things appear smoothly, and if possible warn the audience. Include a 'placeholder' on the screen before the video comes on. This way the audience will be expecting a video and will be concentrating as soon as it starts.
- **Don't have too much on the screen at any one time:** Try not to confuse your audience. Remember that you are the expert, so you may find things very straightforward, where others may not.

There are also a few things to avoid:

- **Don't give mixed messages:** A happy tune or fanfare is unsuitable for introducing poor sales figures!

- **Use the correct colours for warnings or health and safety materials:** If a sign is yellow, use the same colour. Don't change it to make it match your colour scheme!
- **Don't lie:** You can disguise things, but don't try to hide them. If sales figures are poor, try to show how they are better than expected!

Layout

Layout of multimedia products can be very creative, but can also sometimes be very confusing. Produce some sketches of your layout and make mock ups in Microsoft® Word or Adobe® Photoshop®. Test them with a number of people and see what others think.

Colour scheme

It is likely that your multimedia product will be viewed on a computer screen. You will have access to millions of colours, but unfortunately that can mean too much choice!

- Try the colour schemes and effects available in applications such as Microsoft® PowerPoint®, Publisher or FrontPage®. You may be using some other application to develop your product, but you can still 'borrow' the colour schemes from elsewhere.
- Colour can help people to access your material, but it can also cause problems.
- Remember that not all audience members may be able to see colour as well as others, so do not use colour to convey information.

Colour schemes found in Microsoft® PowerPoint®.

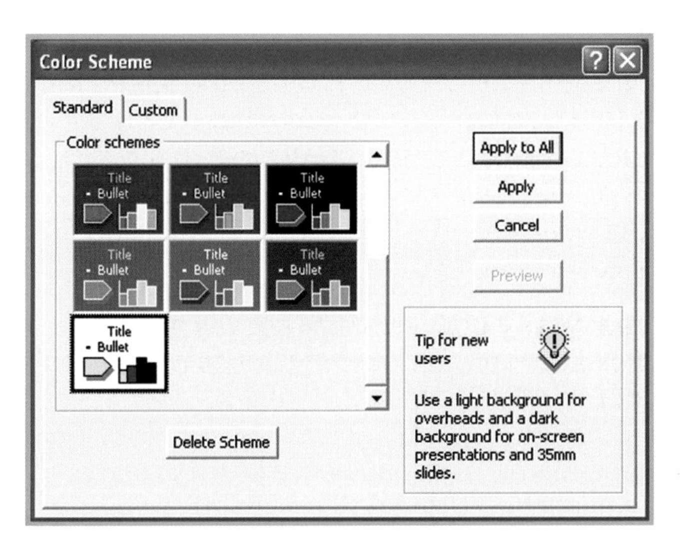

Design templates found in Microsoft® PowerPoint®.

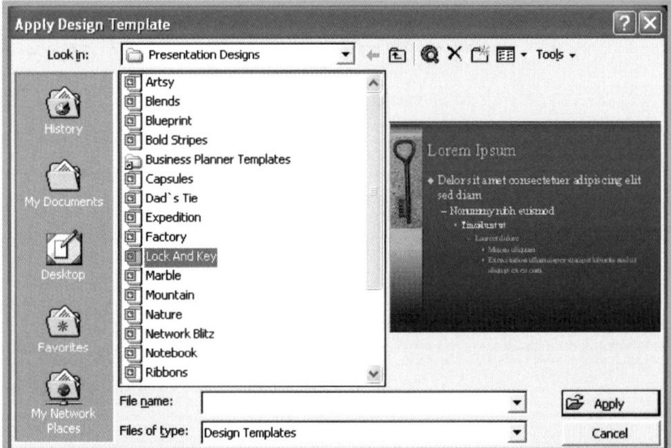

Transferability

If your product is to be transported, how do you intend to do this?

- Transferring data over the Internet is very efficient for small files, but if your final product is more than 2 MB you will need to consider very carefully whether or not the Internet is the most appropriate method.
- If you intend to transfer movies or audio files, you need to check that the end-user's machine will be able to play them. This may mean that you need to include some software or a link to a plug-in that the end user can install.

When you have considered all these aspects and are confident the product can be made to work, you need to prototype some of the real content. Remember that everything may not work at this point, but you need to be confident that they will!

Interactivity

Your prototype should include all of your interactive elements. If you intend to include anything that has actions, you should test them on your prototype.

Hyperlinks are an easy element to add to your work. You just need to highlight an image, section of text, or even a 'hotspot', and make a hyperlink to another file or a web page. However you need to make sure that the hyperlink works when the presentation is running.

Web pages can move or be deleted, so you need to double check that the page you have linked to will be there when you need it. One way of avoiding this problem is to link to the site's home page.

A similar problem may occur with links to files. If you expect a link to open a particular file, make sure that your link points to the right location and that the location remains absolute to the material. The easiest way to ensure that it always works is to save the target file with the multimedia product. If you transfer the product you are then sure that the file moves with it.

If you have built your product with Macromedia® Dreamweaver®, you have the option to test the links. Dreamweaver® can automatically search through your material and wherever it finds a link, it will test it. If there is a file that matches the name in the link it is happy. If, however, it can't find the matching file or the link goes to a website, it will report a 'broken link'. You will then need to change, delete or update the link.

You should also use your prototype to test any other elements of interactivity.

Video or audio can be made to work independently of the audience or they can be made to run in 'players'. If they are running in 'players' they will have play, stop, forward and reverse controls. Make sure that you test them and they work as you intend. If the audio or video file is relatively large, it can take a couple of seconds for a computer to read and start playing the file after the button has been checked. To try to overcome this, make sure you keep file sizes as small as possible. Store the video or audio in the same place as the multimedia product, preferably on the hard disk of the machine that will run the presentation. Obviously, if this is a web-based presentation it will be running on a server, so all you can do is concentrate on file size.

If you have any other interactive elements, you should test each one. Make sure any rollover buttons change as expected, animations run smoothly and other features work together to achieve the result you desire.

Destructive testing

When building products, such as cars, aircraft, etc. manufacturers will test every component until it breaks. They want to know if the product will fail under normal usage (How long will it last? Does it work properly?) and abnormal usage (How stong is it, etc?). These are destructive tests. You need to do the same to your product.

Start your presentation and then try to break it! For example:

- What happens if you click the mouse twenty times?
- Can you make the sound and video overlap?
- Can you make the computer crash or the Internet link fail?

If you get any negative replies to these or similar questions, go back to the drawing board.

You must make sure that your product is robust and reliable. Some of the most prestigious websites in the world have been broken by malicious attacks, but some of them have been brought down by normal use.

You should also make note of how the results of one test affect the other elements of your product. For example, if one sound clip does not play and needs a particular plug-in, will that plug-in need to be installed again or will it be ready for subsequent sound clips?

Compatibility

You need to run your prototype on different platforms and browsers. This will quickly show where alterations in your design should be made. For example Microsoft® Internet Explorer, tends to show some colours differently to Mozilla™ Firefox®.

Microsoft® Internet Explorer

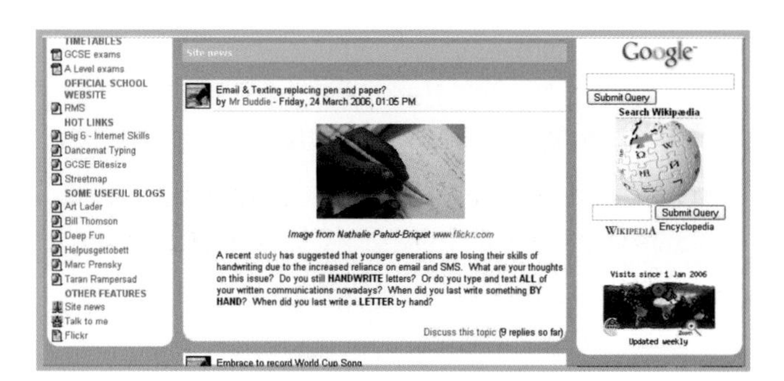

Mozilla™ Firefox

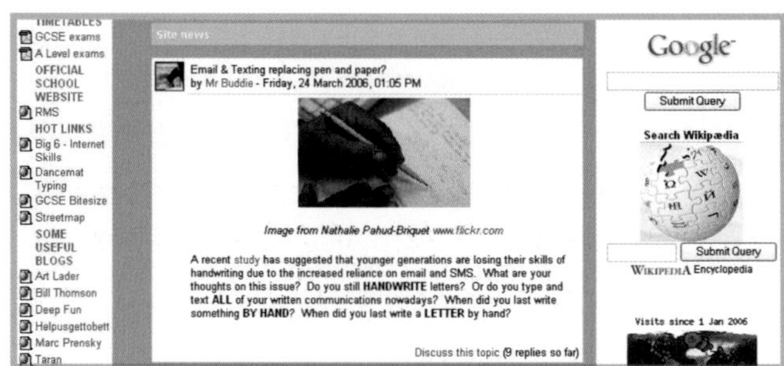

The greatest area of potential difficulty is with the more complex aspects of multimedia production. If you are intending to run video or sound, some browsers will prove more capable than others.

One way around this is to build your product in a specialist application such as Macromedia® Flash®, or Microsoft® PowerPoint®. When you have completed the product, you will need to save it in a format that can be 'played' through a browser.

Macromedia® Flash® allows files to be saved as Shockwave® files. This is a format that plays within a browser, independently of the platform. It does require a plug-in, but it should install automatically. All the file elements are converted into one format, so that the user does not have to worry about the functionality of their machine or browser.

Microsoft® PowerPoint® presentations can be saved in HTML, a common file format that can run on any machine. However, you must check that it works on a variety of machines and browsers, as the application is not specifically designed to produce web-based material.

Help guides

You should consider at an early stage whether your product will be simple enough for any user or whether it will need to offer guidance to some users.

You must ensure that your work complies with the Disability Discrimination Act (DDA) and any other legislation, and it may be that to enable all users to access your product you will need to supply some sort of help guide.

Help should be obvious. Microsoft® Office applications give on-screen help. As your product is multimedia it might be worth developing on-screen helper which need not be as advanced as Microsoft®'s. It might only give a short commentary to get a user started. This could be a particular strength of your product and is definitely something to consider.

Software developers usually employ specialist technical writers to produce help files and documentation.

How to get good marks

✓ **You need to show the examiner that you have tried more than one idea and show evidence of testing. Record the results in a format that is useful and show that you have feedback from a range of users.**

Homework

1. Produce at least three different page layouts for a publication or product and get potential users to choose their favourite and give the reasons for their choice. Record their reasons and show how this information directs your future work.
2. For a multimedia product you have worked on, make a list of all the links and state whether they are absolute or relative. State how you can ensure that the links will always work.
3. Devise a destructive test for a product you have worked on. Produce a short report detailing the test, what happened as you ran the test and what lessons you have learnt from the test.

Distribution

What you will learn in this chapter

You will learn how to produce a multimedia product for a specific audience and develop a runtime version as practice for your e-portfolio.

This chapter explains how to get your multimedia product to the right audience. It will take you through some of the software considerations and other potential problems that you will need to address to successfully distribute your material.

What is distribution?

When marketing a product there are four aspects that need to work together to make it possible for the product to be successful:

- Product management
- Pricing
- Promotion
- Distribution

If any one of these fails, the product will fail.

The DiDA qualification is concerned with the learning of skills and knowledge that will be useful to you when you are seeking employment. Nearly all occupations are involved with supplying a service or product, so it is important that you are confident in dealing with all four of these aspects of marketing.

The management of a project is dealt with in detail in Chapter 8 and the pricing of products is very much down to the individual project as it is often covered in the specification developed from the original project brief. Promotion and distribution used to be clearly defined, but with multimedia products there is a less obvious distinction.

Marketing a product is not easy. Millions of products are designed every year, but very few go on to be successful.

To try to give your product an advantage, you should do your best to promote and distribute it in the most appropriate way.

The distribution of manufactured goods involves haulage by lorries, trains, air freight or other methods. The distribution of e-materials is a little easier – it is certainly quicker!

How to distribute your material

When you have developed your multimedia product you need to consider how you will get it to the correct audience. This usually involves specialist software.

Email

The disk space on an email server allocated to a user (their mailbox) is usually limited to a few megabytes. Any file to be emailed must be less than this (typically 2 mB) to ensure that is received correctly. Multimedia products tend to be large files because of their inherent features. If the file is too large, you will have to compress it to ensure that it can be reliability emailed.

When you have completed the product, you need to find out the size of the file.

1 Navigate to the folder in which it was saved.

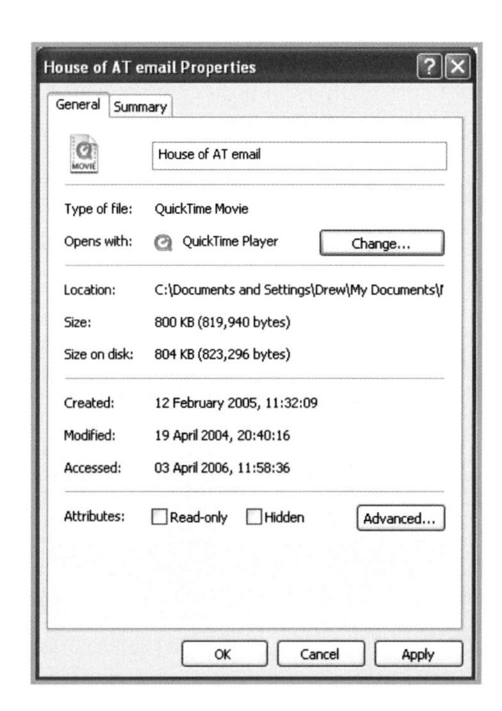

2 Highlight the file.

3 Go to File > Properties.

You may notice that there are two different file sizes. This is because the file on its own is one size, but to be able to open it or work with it the operating system assigns some extra space. When you are distributing the file you must work with the larger file.

If you have been trying to keep the file size under control throughout the project, you will probably find that compressing the file will not make a lot of difference. This is because the files of the individual elements are already compressed. For example, files with the extensions .JPG for images, .WMV for video, .SWF for flash animations, and .MP3 for audio; are all compressed formats.

If your file is still too large you will need to consider a different method of transporting the product.

File transfer protocol

You need a file transfer protocol (FTP) account to upload a website to an ISP's server. The files you upload should not be accessible for modification by others so you need a username and password to log in to the server space. However the server technology does allow public access to the view the files through a browser. Therefore you could distribute your product by sending it to an Internet server. The files could then be viewed through a browser as a website!

The best way to distribute a multimedia product will usually be via a website. Having the product online means that anyone with Internet access can get to see your work.

However, deciding to present your work online is one thing, now you need to think about how!

If your materials are straightforward HTML files, then you should be able to upload them (using FTP) to a server and the links should all remain intact. Internet users will be able to access your material by entering the URL and browsing from the home page.

A multimedia product will often be more complex than a simple website. If you have used video and audio, you will need to ensure that the files are embedded in your product. This can be done using Microsoft® PowerPoint®, or Macromedia® Flash®, but the resulting files tend to be quite large.

To enable your audience to view the files online, you need to decide whether they should be able to view them in their browser or whether they need to download them and view them offline.

When downloading, you need to consider the download speed. A standard dial-up connection can take 40 minutes to download 1 MB. Many Internet users still have dial-up connections. Even those with broadband may have problems downloading larger files, so it may be appropriate to view your material 'live'.

If they are to be viewed in a the browser, you need to think about the 'run-time'. This is the order that the viewer will see things:

- Microsoft® PowerPoint® generates a sequence of slides. The viewer usually starts with slide one and moves through to slide two, slide three, and so on. If you are using Microsoft® PowerPoint®, try to avoid putting too many elements on the first couple of slides, so the browser can download the remaining slides while the first slides are being shown. If you have added video, having two or three slides at the start, without video, gives the software time to start downloading the video file, so that when the viewer gets to the slide, the video is ready to play.
- Macromedia® Flash® has a particular way of displaying materials. As you make the SWF file ready for uploading you will be asked which order you would like the file to download (top first is the default setting as most Flash® developers work from top to bottom on their timelines). This means that the user will get to see the files as they appear on the timeline. As the file is opened by the browser it will start to display the top items on the timeline. As the viewer watches the first few parts, the remainder are downloaded and are then waiting, ready for the viewer to get there.

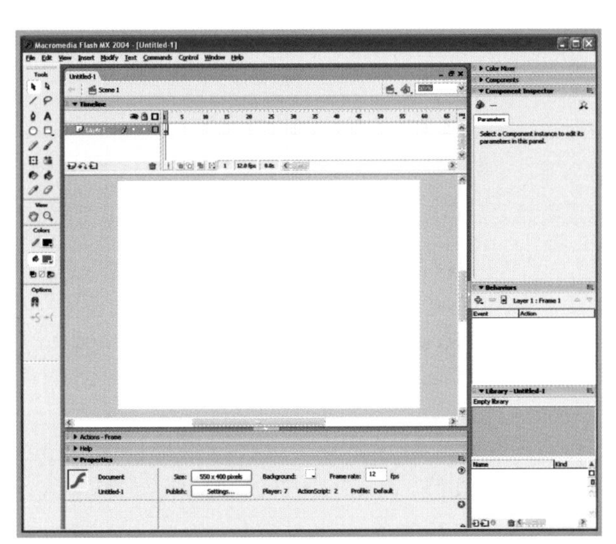
Flash® interface, showing timeline.

Both of these methods allow the browser application to manage how the file is downloaded and therefore let the user see the product as you intended them to.

If the user is expected to download the files and view them offline, you will need to include instructions on the download page:

If you wish to view this product offline, you will need to download it and save it in a folder on your computer.

When the download is complete you will be prompted to open the file, if you do so it will generate a browser window and start to play the file.

You should also explain to the user that any links to external websites will only work if they have an active Internet connection.

Other distribution methods

In industry, multimedia products are often distributed on recordable media, such as CD, DVD, memory sticks or flash memory.

The DiDA qualification requires your e-portfolio to be less than 15 MB, so it is unlikely that any multimedia product you develop will be larger. However, you should still consider how you could distribute a larger product.

Format	Capacity	Advantages	Disadvantages
CD	700 MB	• plenty of space; cheap (less than 10p each); • easy to package	• specialised writer required; • often too large for content so space is wasted; • can be scatched.
DVD	4.7 GB	• massive capacity; • cheap (less than 10p each); • easy to package.	• see CD disadvantages, and • confusing variety of formats.
USB memory stick	Up to 1 GB	• available in a variety of sizes; • highly portable; • they look good!	• expensive (>£10 for 16 MB); • difficult to label.
Flash memory	Up to 1 GB	• available in a variety of sizes; • fast; • easy to use.	• see memory disadvantages, and • card reader required.

If you are distributing a self-contained product, you may wish to include software to ensure that the user can view your product. For example, if your product requires Microsoft® Internet Explorer v5 or later, you may wish to include a copy so that the end-user can install it on their own machine, or if you are showing a Microsoft® Excel® spreadsheet, you may wish to include a 'viewer', to allow a user to be able to use the spreadsheet, even if they do not have the application.

Applications such as browsers, plug-ins, specialist players and other materials can often be found on free DVDs available with computer magazines.

Targeting users

Getting people to come and look at your multimedia product is the most difficult bit of marketing. Most products have a specific audience in mind: home shoppers, male youths, female youths, vegetarians, BMW owners and so on. You should be able to work out from the specification who your intended audience will be. You then need to think about the best way of getting your product to them.

How can you advertise your product?

Linking your product to a website will not get people to suddenly flock there to see it! You need to let people know that it is there. There are many ways of doing this. You should consider them all, and then choose the most appropriate.

TV advertising
This is very expensive, but at peak times there can be up to 20 miliion viewers. Some of the less watched channels run advertising relatively cheaply, but then there are fewer viewers.

Radio advertising
A national advertisement is expensive, but most independent radio stations are 'local', so the advertisement costs are lower, but the advert is only broadcast to a local area.

Newspaper and magazine advertising
National advertisements can be very expensive, but as with local radio, the local newspapers or magazines are less expensive to advertise in, but they only service a particular area.

Telephone advertising
This is called telesales. Some people refuse to accept telesales calls. Most people consider them a nuisance at home, however many organisations use this method to deal with other organisations.

Postal advertising
Sending letters to the general public, regardless of the content, is often considered to be junk mail. If you use post, make sure you can write to specific people, using their name.

Web-based advertising
Pop-up advertisements are annoying to most users, and can be blocked by anti-virus software. They are less successful than they were in the past. Placing an advert on a popular website can be relatively cheap, but again can prove counterproductive, as some users will be annoyed by it!

Emailing potential users
General emails to a wide number of recipients are often considered to be **spam** and so will be blocked by some servers and email software.

Appearing in search engines
Over time, nearly all websites are hunted down by search engine 'spy bots', and so will eventually appear in their listings, however if you want to get your site listed quickly and appear as the top results when someone uses particular keywords, you will need to enter the details in a listing application or pay the search engine provider.

Word of mouth or recommendation
This is by far the best form of advertising, and it's free! To be able to benefit from this you need to have a very good product, which meets the needs of the users.

You will probably be aiming your product at a relatively small group of people, but that does not mean that you need not consider a variety of ways of targeting your users.

Your work must reflect that you have looked at different methods and chosen a particular strategy for a number of reasons.

How to get good marks

✓ You need to create runtime versions of your products, showing that you have used appropriate specialised applications to produce your multimedia product.

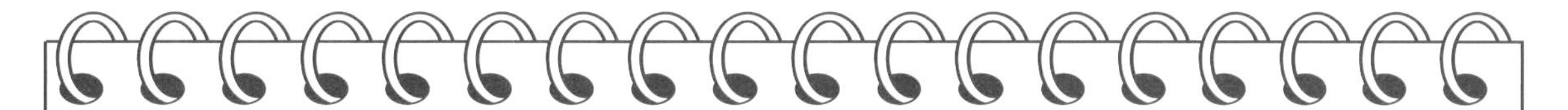

Homework

1 Carry out a test to find out how much disk space can be saved using a compression application, such as WinZip®. Note the file size before and after compressing, then work out the difference between the two values. Which file types can be compressed the most?

2 Collect a set of applications or players and burn them onto a CD or DVD, ready for installation, so that a user will be able to view the following file formats on any suitable machine:

- PDF
- HTML
- MOV
- DOC
- XLS

3 Produce a report explaining why you would choose a particular method of targeting your audience for the following products:

- A new washing powder.
- A second-hand car.
- A new motorway service station.
- A discount week at a sports centre.

Presenting multimedia products in an e-portfolio

What you will learn in this chapter

You will learn how to make it easy for your examiner and your teacher to find saved work, and make sure that your e-portfolio is efficient and easy to use.

Each DiDA unit is assessed via a summative project brief (SPB) in which pupils bring together the knowledge, skills and understanding they have acquired throughout the unit into one substantial piece of work more on the SPB can be found in Chapter 17. This project is set by the exam board and sent to the pupils at the place they are taking the DiDA course. The project is an important and sizeable piece of work, which must be done in school, college or whatever centre in which the learning is taking place. It is recommended that a minimum of 30 hours should be allowed for each project.

All the work is saved and when finished sent to an examiner over the Internet. This work is marked by the teacher/lecturer and externally moderated. Pupils always complete the SPB project toward the end of the course. When your e-portfolio is assessed the examiner will want to find your work easily and not spend ages looking for work that has been poorly labelled or hidden in some obscure folder that is not properly named. This will, of course, lose you marks as hidden or difficult to find work may be lost or assumed not done.

You should usually save work in folders. These can be made easily and if named and saved logically will make it easy for you, your teachers and others to locate different parts of your course work.

Making folders

When making a new folder be careful when naming the folder so that it does not cause confusion. Make sure the name is relevant to the folder's contents. By making folders with distinctive, relevant names things become easier to access. Having to spend time looking for certain documents or files should not be a problem. You can make subfolders, but sometimes it is more convenient to wait until you have more to add to the folder before you create it.

When creating a project, get into the habit of creating folders to organise all the files associated with it.

Making a folder in Microsoft Windows XP

One of the many methods that you can use to make a folder is shown below.

The folder tree is a visual representation of the locations where files are stored. So you can use this to navigate to the files you want to find. To do this you need to visualise the folder tree as a tree growing from left to right. Clicking on the small plus sign + will open a folder which contains more folders or files. If you then click on the small minus sign −, the folder will collapse and return you to a higher level of the tree.

1 Click on Start and then click on My Documents.
2 Click on 'make a new folder' (at the top of the left hand panel of the My Documents screen).

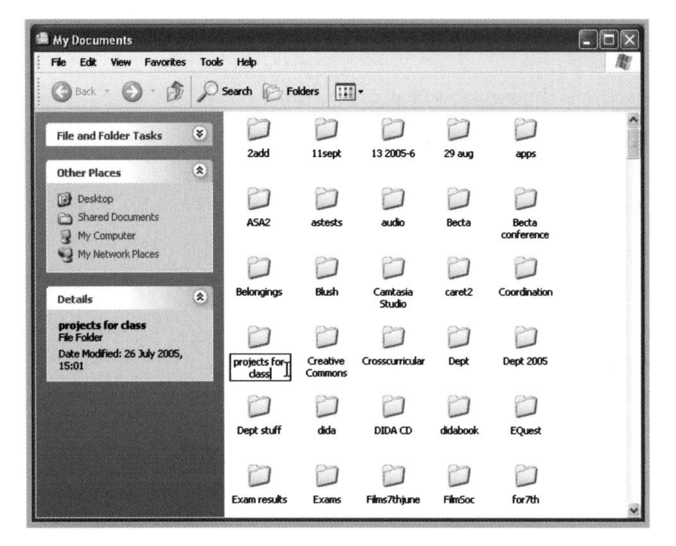

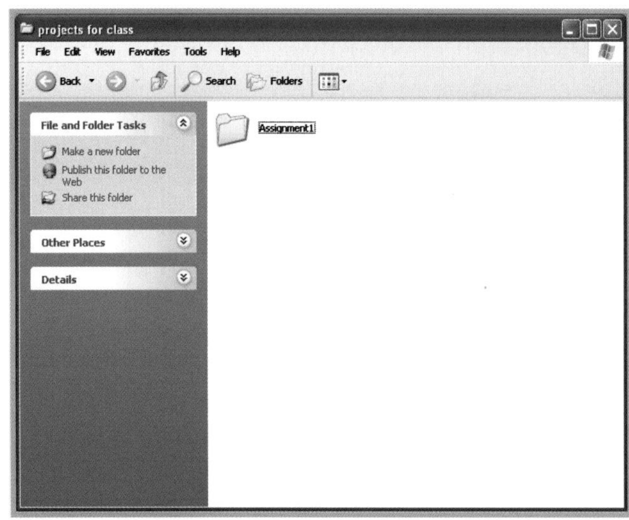

3 A 'New Folder' icon will appear in the file list. (Note that this list may appear either as a folder tree or as a display of icons representing files and folders depending how Microsoft® Windows® is configured.) The text will be colour highlighted. Type in the name of your new folder. Press the Enter key or click in a 'neutral zone' off to the side to accept the name of the folder. If you make a mistake, right click on the folder, choose Rename from the submenu, and type in the correct name.

4 You can make sub-folders within your folders. Double click on the new folder. It will be empty. Click on the 'Make a new folder' command once again and another new folder will appear. This allows you to create a main folder for your files and then organise related files into subfolders. For example, you will have folders for pictures, sound, movie clips and so on.

5 If you don't want a folder, click on it to select it, and then press the delete key. Deleting a folder will delete all its contents, including subfolders and their contents, so move any content before deleting!

Project planning

Marks are awarded for project planning, selecting and capturing information, evaluation and the e-portfolio, in addition to the specific requirements of each of the four units.

It is easier to work through a plan of your project than to jump from one unplanned action to another. The planning, if done well, should speed you along, give you clear aims and objectives, help you judge where you are in the whole project, and help maximise the marks you will be awarded.

When you have finished your project, review your work. You should evaluate the project outcomes and your own performance, including feedback from others and suggestions for improvement.

Table of contents

A Table of Contents (TOC) is used to show the topics that are contained within a document. It has a vital role to play in showing the user how to find the location of items within the document.

There are several different way to create a TOC using Microsoft Word®. You need to decide on how to present your TOC in terms of its format and the inclusion of page numbers. Once you have established these basics, you can then decide where you want to place the TOC within your document or website. It may be that as a student you want the moderator to easily identify the location of your work whilst giving an overview of the project. In such a scenario a TOC would be of great help as it could be included on the homepage of the website.

A paper-based TOC consists of some basic elements:

- Lists of topics
- Leader lines (dotted lines that fill the space left by the tab characters)
- Page numbers.

However, it is not necessary to the last two items on web-based documents as hyperlinks will take the user to the relevant section.

In order to create a TOC, you need to highlight the text that will appear as an entry in the TOC. This portion of text should be as short as possible, but still needs to describe the content appropriately. Once the text is highlighted it should be inserted into the TOC.

Microsoft Word® uses Heading Styles to assign specific formats to headings. It uses nine different preformatted styles and if Microsoft Word® can identify which level of heading applies to a TOC entry, then it will automatically apply that style of formatting to the text.

This way of formatting can be seen by highlighting the text within a document. Then by choosing a format from the list that appears when you choose the Table of Contents Options dialogue box that format will be applied.

TC fields

Table entry fields (TC fields) are special codes that include the letters TC with { } placed around them. A TC code can be used to insert the text contained with in it directly into a TOC.

TC fields can also be used to customise your TOC by adding a switch (\) in the TC field.

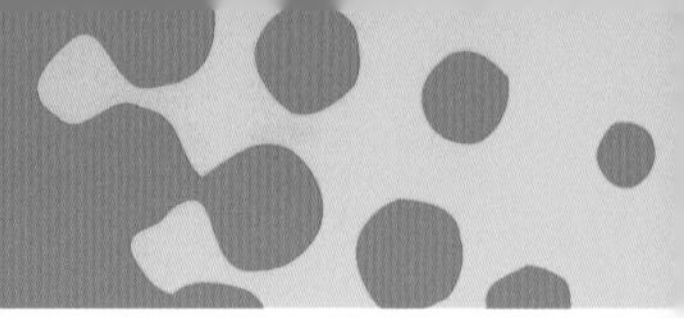

To mark a TOC entry with a TC field, highlight your chosen text and press ALT+SHIFT+O. This displays the Mark Table of Contents Entry dialogue box.

Saving your project

Your e-portfolio is limited to 16 MB, so how you save the file is very important as this will determine how much space you use up and how long something takes to appear on the screen. Simply saving a file is not enough. You need to think about file types and why they are used. Each file type has its own particular purpose and place. If you are using a lot of photographs, movies or images they can quickly use up your allocated disk space. If you are saving images think about the file types below.

- Joint Photographic Experts Group (JPEG) images are compressed using an algorithm to discard data to create an image whilst not identical to the original, is nonetheless very close to the original. Such images tend to be compressed by a ratio of somewhere between 5:1 and 15:1.
- The Tagged-Image File Format (TIFF) is used when scanning images and documents and is most useful when transferring documents between applications.
- Encapsulated Postscript file (EPS) format is used with publishing software packages and image manipulation packages such as Adobe Photoshop®. One disadvantage is that these files can be very large.
- Bitmap (BMP) is the format that files are saved as from Microsoft Paint®. The backgrounds or wallpaper used for the main Microsoft Windows® screen are usually in this format. A limitation of this format is that the colour palette cannot be changed so when files are converted to BMP format, the colours may alter from their original state.
- Graphics Interchange Format (GIF) files are highly compressed and hence they can be transferred quickly by email or other electronic means.

Viewing work on different computers

PCs and Apple computers display images in slightly different ways resulting in text being larger on a PC whilst ant-aliasing on the Apple platform gives a smoother appearance to text.

While paper-based publications can be designed using many fonts and styles, web designers have more limitations placed upon them. This is because a user can only utilise fonts that are installed on their computer. Any fonts which are used and which are not installed on a computer will be displayed as default fonts and may not convey the image that the designer intended. If a designer absolutely wants to make use of an unusual font then it has to be displayed as an image.

Computer systems can present the same information to the user in many different ways. It comes as no surprise then that this has a range of problems associated with it. The designer has to bear this in mind.

User preferences

As well as the restrictions posed by the designer's computer, it is important to remember that the users computionary place restrictions on the context. For example, screen resolution may be altered and this can have a major effect on the way a web page looks on the screen. It is also possible to use different web browsers which will present web pages in slightly different ways.

A good web designer will take account of this when designing a site because they can never be sure what a user will be using to look at a web site.

It is sometimes possible for users to alter the way a screen is displayed from within the browser itself. The most obvious example of this is the way text is displayed to support people with visual impairment. Microsoft Internet Explorer® allows text to be displayed in five different sizes whilst Mozilla Firefox gives more flexibility through the use of Ctrl+ and Ctrl– .

Webpages assess the way a page has been set up and then render it as required which accounts for the various ways it can be displayed by different browsers.

It is sometimes the case that pages do not get displayed properly, but this can be as much the fault of the browser as it is of the webpage. This only makes it more important for designers to test their web pages on not only a range of browsers, but a range of versions of browsers.

It is hard to build a website that will have been tested on all available browsers – you would need a large team of designers to be able to do this efficiently. It may be sufficient to test the way it looks on the latest two versions of two or three of the most commonly used browsers.

Don't forget that unlike paper-based items, original web pages can always be redesigned if necessary to take account of new browser features.

Testing your e-portfolio

When you have saved work into your e-portfolio check that your folders and files:

- are easy to follow;
- work in a range of browsers;

- have no mistakes, spelling or others;
- function how you would like them to;
- are clear and easy to navigate through.

Ask a friend to run through your folders, to find a file, check that they can open a file and that it is easy to do so. Ask them to review the content and say what they felt about the search. Was it easy or could it be improved?

Disaster recovery

What could be the worst thing that could happen to you work? Have you backed up your e-portfolio files? You will have spent a lot of time working on your solution to the SPB. Your teachers or lecturer will have agreed that the work was yours and the examiner will then be waiting for it to check it. What would you do if you lost your work because the network collapsed, your computer got stolen, or the room was gutted by fire? Disaster recovery systems should prevent any problems arising if any of the above happened.

Here is what you could do:

- Regularly back-up your work on a CD or two.
- Take a copy home or to a different location away from where your work is normally saved on network or hard drives.
- Send your work to yourself by email or to some other people (let them know first!). You may have to do this a bit at a time as the files might be huge.
- Ask the network manager if your work is automatically backed up.

Summing up

Navigation

When you are ready to start building the e-portfolio you need to arrange it in such a way that your teacher and the moderator can find everything easily.

This means that you will need to develop some sort of file structure, with a *contents* document containing links showing where each piece of information can be found.

Ideally you will have a variety of documents or different files that are grouped together as a website. Each file is linked to a 'homepage' and can be accessed from there.

The Home page

The main user interface of your e-portfolio will be the Home page – so make it a good one!

Try to make your Home page as interesting and professional looking as you can. Use all the tricks you have picked up through the course to impress your teacher and the moderator.

The Home page should be of a standard file type, such as HTML, and will contain a number of important details, as well as carrying a short description of the files or documents:

Homepage details:

Name

Candidate Number

Centre Number

Project title

Date of submission

Group name/Teacher name

A well-designed Home page is a great advantage in your e-portfolio. You could build a 'smart' page, with rollover buttons and interactive messages, or keep it plain and simple. Whatever you decide, you should introduce the reader to what they are about to see. Try welcoming the reader:

Welcome to my e-portfolio. Over the last three months I have been working on a project: **Producing material for the launch of a new restaurant**. The links below will take you to a variety of documents which detail the processes I worked through and the outcomes I generated while trying to satisfy the various requirements of the project brief.

Although each link has been tested before submission, if anything does not work, each folder contains an index with a set of links to each file within the folder. If there are any further problems, you can email me at dave@e-portfolio.co.uk

Images
Here you will find 10 images of material I have produced
1 A design for a logo (to appear on the headed notepaper and menus).
2 Four versions of the headed paper and menu designs.
3 A postcard to be sent to prospective customers.
4 A poster to appear in a local public centre.
5 A handout to be distributed in a shopping centre.
6 An on-screen presentation.

Project Planning
Here you will find my documents related to how I planned this project
1 Initial plan.
2 Timesheet.
3 Revised plan.
4 Quality review record.
5 Diary.

Information Sources
Here you will find details about my sources of infromation for this project
1 Bookmarks (with short descriptions of each site).
2 Bibliography (details of all the books and magazines I have used).
3 Image library (a list of images and image galleries I have visited).

Data
Here are the results of my research carried out by a survey and interviews.
1 Questionnaire.
2 Results.
3 Interview script.
4 Interview 1 transcript.
5 Interview 2 transcript.

Information
Here are files that show how the main aim of this project was to show how I met this project's main aim of producing material and communicating information about the new restaurant.
1 Document templates.
2 Website designs.
3 On-screen presentation of a sample menu and opening times.
4 Prices.

Review
Here are some of the most important documents. They have been developed from my research, tests and evaluation.
1 Original specification.
2 Quality review record.
3 Testimonials from the restaurant staff.
4 Overall evaluation document.
5 Proposals for future work.

The Home page must be stored in your root folder and be clearly identified. Give it a name that will mean something to those that view it. You should use something like your initials, candidate number and unit number: for example DP_01234_Unit01.htm. Within the root folder you should set up a folder structure that enables easy navigation.

Possible structures

The qualification specification contains details on the marks available for each piece of work. However to help your teacher and anyone else who needs to assess your work, Edexcel have broken the marks down into six categories.

Category	Available Marks
Plan and manage the project	5
Select and capture information from a variety of sources	7
Collate and analyse data to produce information	7
Present and communicate information	9
Present evidence in an e-portfolio	9
Review the project	5

You can use variations on these six categories as your folder titles or you may wish to break your e-portfolio into other sections related to the evidence type each folder contains: images, spreadsheets, databases, paper publications, on-screen publications, etc.

Each folder should contain an index page with links to each file to make navigation very clear.

Each file must be clearly titled and saved in a format that can be easily accessed through a fifth generation browser.

Bear in mind that the marks mentioned above for the e-portfolio are awarded across all of the material you submit, so you do not need a section or folder for that aspect.

The audience

You should bear in mind the *intended audience* at all times: your teacher, the assessor and the moderator. These people may not know you or your work, so the only chance you have of letting them see your best and most applicable material is by very clear navigation around your e-portfolio.

A list of acceptable file formats will be issued by Edexcel, with the project brief. These are likely to be pdf for paper-based publications, jpg or png for images, html for on-screen publications and swf (Flash® movie) for presentations, but this may be revised to take account of future developments.

To enable you to work with these file types you will need access to conversion applications. Your teacher will provide guidance on what is available to you in school. For home use, you should have access to:

- a compression utility, such as WinZip® or Allume Stuffit®;
- a pdf generator, such as novaPDF;
- a Microsoft® PowerPoint® to Macromedia® Flash® converter, such as FlashPoint;
- an Image converter, available with most image-manipulation software, (e.g. Adobe® Photoshop®, etc).

The total file size of the e-portfolio may be around 15 MB, which is not very big. You must work out exactly what you need to include and then keep a close check on the size of the folder containing the e-portfolio. You may be able to keep file sizes lower by using optimisation or compression formats. You must present your e-portfolio content in a format appropriate for viewing at a resolution of 1024 × 768 pixels.

As with all websites, you must ensure that your material follows the rules on accessibility: Where possible you should:

- provide alternatives to auditory or visual content;
- avoid using colour for navigation;
- use simple, appropriate language;
- use tables;
- design your materials to work on any platform;
- provide clear navigation systems.

Before you submit your e-portfolio, test it! Carry out tests on a number of different machines, running different browsers. Remember, just because it works once does not mean it is reliable. Your qualification is resting on the reliability of your e-portfolio.

Clear identification

Each file must have a distinct title and a short explanation of its content. Remember, you will know what everything is, but will someone who looks at your work for the first time?

It may be difficult to add comments to images, however embedding the image in a DTP document and then adding your comment prior to converting the document to pdf may be a solution.

There may be similar problems with other filetypes. If you experience difficulties, it might be worth generating a short page of text as the main link, and then using a hyperlink from that to the file:

You will see from the example that it is important to make the reader aware of any legal issues that you have had to take into account during your work. Any data that contains names and addresses comes under the Data Protection Act 1998, so unless you have permission, you must not pass these details to others.

There may also be copyright or other issues that you have had to deal with. If you have scanned an image, have you asked permission? If you have used some text from a book, have you recorded where it came from and acknowledged it in your work? Do you have permission to use photographs of people in your images folder?

Answers to these questions should be included in the commentary that accompanies each file.

Submission

- By the time you submit your e-portfolio, you should have a complete package of files and folders, all clearly linked and accessed through a professional looking Home page.
- You will have tested it on a number of different machines and asked others to test it for you.
- You will have incorporated the suggestions that the testers made, to improve your work.
- You will have checked every link and removed any that do not work.
- You will have checked that your e-portfolio contains everything that you need it to contain.

- You have checked the total file size and it is below the 16 MB limit set by the Awarding Body.
- You have checked that every file opens quickly and displays as expected.

Now comes the time to submit your work.

Although your work may be submitted to the moderator electronically using a secure Internet link, it is likely that you will be asked to submit a separate copy to your teacher for marking.

You should decide on the most appropriate method of submitting your e-portfolio. There are a number of options shown in the table on page 63). You should check with your teacher to make sure that the format you choose is acceptable.

Although there are no marks for the submission aspect of your e-portfolio, how you do it can add a feeling of 'professionalism' to your work. Don't be afraid to make it look good!

Finally

Proofread the content on-screen. Check spelling but do not rely on your spell checker as these will often put in a word that is similar to the one you misspelled but might still be the wrong word. For example, bite is a word you may have misspelled for byte: the checker will leave it!

Check everything works well and that what you intended to produce is actually functioning.

Do not be critical of your work before you show it to others. Testing usually includes showing your work to people, not just yourself. Evaluate your work and try to be positive, think about what went well and what you would change if you could do it again. Let others know what you enjoyed and which parts you are pleased about. Talk about your achievements.

Post-mortem

At the end of any project, you should allocate time to review the information on both the work itself and the management of that work.

How do I get good marks?

✓ You need to make sure that it is easy for your assessor and moderator to find your work. To help, you need to have a homepage, table of contents, references and copyright information. If you have done the work, but the examiner cannot find it, you have wasted your time!

Homework

1 Experiment with some designs for the Home page. Personalise it, but make sure it is looks good and works well.

2 Ensure that you have access to the appropriate file conversion applications.

3 Check the accessibility rules at www.w3.org

4 Before you submit any work get someone or several people to go through it. This is called testing and this may highlight any problems. If possible, do this away from the classroom, perhaps at home. Evaluate your results.

Project planning

What you will learn in this chapter

You will learn how to plan a project using the same strategies as a professional project manger. You will learn when and how to review your work and the importance of allocating time.

This chapter will describe how to manage a project as complex as a multimedia product. It will also explain why certain things should be considered at particular points in the development of your products.

If you do not plan your work from the outset difficulties will arise straight away. Project planning will help you to keep moving forward as you work. You will find you are not waiting for things or suddenly surprised by a problem that has loomed up! In industry project planning is always done before a large project begins. This is because it will help people decide what and who they need and when they are needed.

Each unit of the DiDA is assessed via a summative project in which pupils bring together the knowledge, skills and understanding they have acquired throughout the unit into one substantial piece of work. This is marked by the teacher and externally moderated. Pupils usually complete the project toward the end of the course.

The project is an important and sizeable piece of work, which must be done in school or college. It is recommended you spend a minimum of 30 hours on this project.

Why plan?

It is easier to work through a plan of your project than to jump from one unplanned action to another. The planning, if done well, should speed you along, give you clear aims and objectives, help you judge where you are in the whole project, and help maximise the marks you will be awarded.

At the outset you will be given a set of intended objectives. You need to make sure that you clearly understand these objectives because they will be used to assess how successful your multimedia product has been.

You also need to be confident that your project has a purpose. You need to know where you are going and why!

'Great minds have purposes, others have wishes' – Irving Washington

Most of your multimedia products will have at least two purposes:

- To satisfy the client's needs.
- To help you to gain your DiDA qualification.

These two purposes should complement each other. The nearer you get to satisfying the client's needs, the nearer you will be to achieving the marks needed to get your DiDA.

Professional project planners often use specialist applications to help them manage the tasks that build together to make up a project. Some of these may be available to you, for example:

- Project planning software, such as Microsoft® Project are complex arrangements of spreadsheets, databases and charting systems. If you have access to this type of application it is worth taking a look, as you may wish to use it for your work, but it may take you a little while to get to grips with it.
- A slightly more straightforward system is mindmapping. This is a method of noting everything you can think of that is related to the project, then adding detail to every point. The individual aspects are then drawn together as a set of tasks that can be sequenced. There is a range of these applications available to download. Try searching for 'mindmapping'.

Both of these software tools are designed to help you manage your workload. Even if you decide to project manage another way, you will still need to meet the needs of your client, so you must develop a specification.

The specification

A specification is:

- the definition of your project;
- a statement of the problem.

It is not the solution. So that you don't waste time working on the wrong project, you need to write a specification. This is a definition of what is required, by when. This must be agreed by all involved. There are no short cuts to this. If you fail to spend the time initially, it will cost you far more later on.

A written specification has several benefits:

- The clarity will reveal misunderstandings.
- The completeness will remove contradictory assumptions.
- The rigour of analysing problems will expose technical and practical details.
- It forces you to read and think about the details of what you need to do!

To enable you to successfully manage and develop a multimedia project, you need to consider the following questions, as part of your specification:

- **What is my product for?** You need to be sure that you know why a multimedia product is considered to be an appropriate solution.
- **Who is the intended audience?** You need to know who the product is aimed at. This includes: age, gender, personal interests, approximate income, family status and any other more relevant information.
- **What multimedia components are required or are available to me?** Your client may have stated that they want a certain number of images or a particular amount of video. You need to make sure that you include that detail so that you can tick off when you have done it.
- **What are the technical requirements?** You need to be aware of any limitations, such as colours, screen resolutions or size, as well as any restrictions on file types.
- **Timing?** Before you start anything, check what the timescales will be.

Identifying errors in a specification involves the following:

- The context: too tight a focus on the context and not the substance of the project itself can cause problems.

- Requirements: these should be discussed at the outset and a failure to do so can jeopardise the whole project.
- Plan your time wisely because if you fail to do so it is likely that you will spend too long on sections that are irrelevant or which will gain you few marks.
- Set realistic deadlines: without these you cannot tell how your project is progressing.
- Be certain that you have access to the resources you need. For example, it is no good hoping to use a scanner if you do not have access to one.
- Set measurable targets – otherwise you cannot monitor your progress.
- Set milestones that can clearly identify the things you should complete and the date by which they should be completed.

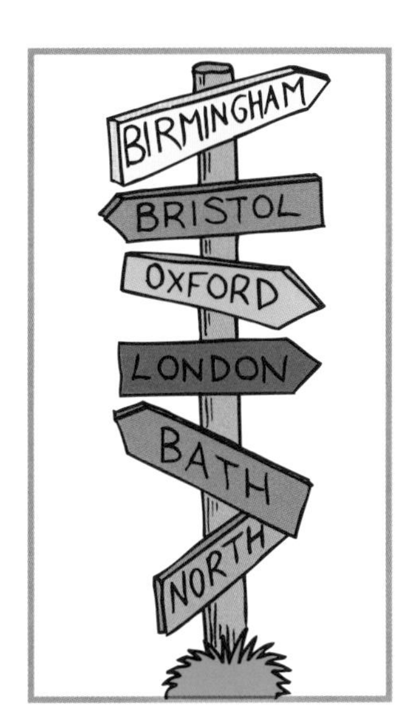

If you set your milestones carefully, you will find that they will aid you in the planning of your project. You can use these milestones to help you to identify the critical points of your project.

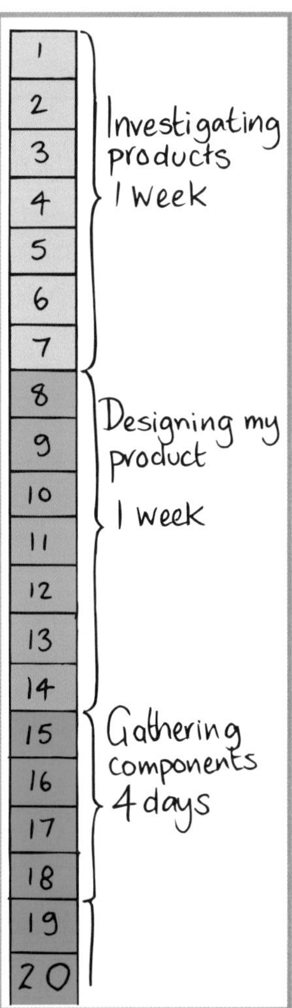

You can set your milestones by identifying the date by which your project must be submitted, then deciding which subtasks have to be carried out. Once you have done this you can then set sensible dates by which each subtask should be completed. These will be your milestones.

Timeline

A timeline is a linear (line) plan on which you can plan where you want to be each week, month or even day. It's a very visual guide that will help you decide if everything is running to schedule. You can highlight important points on the timeline and schedule checks and feedback dates. You can leave space on the timeline so that you can make notes or plan visits or meetings.

Once you have developed your timeline, print it out and stick it on a wall so that you can refer to it regularly.

Make sure that you not only plan your time, but that you record on your plan how well you are getting on. A plan is supposed to be a working document. It should change as you work through your project to reflect the differences between what you have estimated and what

time is actually needed. The revisions to the plan will help you to estimate more accurately in the future.

You will have a number of stages in your project. You should record your work at each stage and use this information to form your project evaluation. There is more detail on evaluations in the next chapter.

Common problems

There are a range of difficulties that may present themselves as you work through your project, but careful and accurate planning can enable you to avoid many of them. Some of the most common are listed below, make sure you plan around them.

- Resources are not always available.
- Extra tasks pop up during the course of the project.
- What happens most frequently is that the initial estimate of the time required to perform a certain task is wrong!

What have I got to do?

This is task allocation (what do I do when?)
Make a list of things you need to do. You should now allocate the tasks in a sensible order on your time line: think about what you need to do first, then second and so on.

Task allocation is not simply a case of allocating the tasks on your lists to the time you have available. Think about what you can do now and what you need to prepare for and do later.

Consider what tasks you have to carry out and allocate sufficient time to the tasks (give yourself a little extra time whenever possible).

Developing a multimedia product is likely to entail using a number of the more advanced software applications. Some of these can look very complicated, and you may need time to get to grips with them. Take a look at the Help files that come with the software, they sometimes contain tutorials. If you feel daunted, break the tasks into smaller units or look on the Internet for guidance.

Sometimes tasks can be grouped and allocated together. For instance, some tasks which seem to be independent may benefit from being done together since they use common ideas. If you are converting images from RAW format (straight from a digital camera) into JEPG format (ready for adding to your product), it may be worth investigating 'batch processing', where the software carries out repeated actions automatically, without your intervention.

Always plan ahead. Do not wait until you need something. Make sure it's there ready for you when you need it.

Testing and quality

You need to make sure that each time you think you have completed a task, you have it checked by someone else. This is a crucial part of the process as it is only by receiving feedback from your target audience that you can see whether you are really achieving your aims. It is wise to ask several people to look at what you have produced. The reason for this is simple – one person may give you their own, objective, opinion whereas when you receive comments from a group of people you can come to a consensus of opinion. This in turn will lead to more meaningful recommendations as to how your work could be improved.

Obtaining feedback can be a time-consuming process. Therefore make sure you plan for this in your timeline. You must not assume that a person will be able to give you feedback as soon as you ask for it or at a time that suits you. You may have to wait several days until the person can let you know what they think of your output.

It is also important that you test that your output works efficiently. You must build a testing schedule into your plan too.

Fighting for time

You need to stick to your schedule as closely as you can. Try to resist any temptations to be distracted by new challenges as they arise. If they are important you can always tackle those once you have completed what you originally set out to do. If other individuals set a deadline that you find hard to meet, you must explain this in your report. Sometimes you will have to meet such deadlines, but if you have good reasons why they could not be met, then it is good to state these as justification.

Fitness for purpose

You will find that you work well as long as you remain motivated. As with any extended task, you can easily become disinterested as the task may seem never-ending. However, if you constantly remind yourself of what you are working towards then you will gain all the motivation you require. This is also where the use of milestones can help as each time you reach a milestone, you will get a sense of achievement that should spur you on to your next milestone.

You may feel pulled in different directions as you have to prioritise the elements that are most important to you. You may need to combine the skills you have by using a variety of packages to complete a particular task. To this end it may be more efficient to look for tools that do the same job within the one program e.g. Why draw images in a graphics program when you could use the drawing tools within Microsoft® Powerpoint®?

Planning for error

A common mistake to assume that everything will work perfectly first time. You need to be aware that this is rarely the case, and more often than not the desired outcome is not what is produced in the first instance. It is important that you incorporate time for things to go wrong in your timeline. If something disastrous happens you will then have some breathing space rather than having to panic.

The creation of multimedia output is a time-consuming process. So although you hope that the worst will never happen, planning for things to go wrong is one of the things you must do if your project is to be a success.

Sometimes it takes another person to notice something that we ourselves cannot see. This can happen because you are too involved in a project and cannot 'see the wood for the trees'. So it is useful to ask your classmates to look at what you are planning and suggest potential pitfalls and areas of concern. You can then revise your plan in the light of the feedback you receive.

By following this advice you can highlight the elements of your plan which provide the greatest potential for things to wrong. Then you can plan to spend a little longer on these areas to ensure that everything has gone according to plan.

How to get good marks

- ✓ **Keep referring back to your plan and your objectives to see if you met them.**
- ✓ **Evaluate your objectives, whether you achieved them or not, and why.**
- ✓ **Get feedback from others and show what you think about what they have said and whether they were right.**
- ✓ **Consider the whole process of your project from beginning to end.**
- ✓ **Make some suggestions for improvements to your project and explain them.**

What will lose me marks?

- ✗ **Not referring to your plan.**
- ✗ **Reviewing your objectives merely as checklist saying yes or no to each objective.**
- ✗ **Not mentioning feedback from others.**
- ✗ **Just saying what you did without explaining how or why.**
- ✗ **Not making valid suggestions for improvements**

Homework

1 Produce a single page breakdown of the project you are working on. This should be presented in an interesting way.

2 Put together a timeline for an activity you are confident with, such as making a cup of coffee or tea. Are there any activities that could be done while others are being carried out, such as adding sugar while the kettle is heating the water?

3 Produce a template for an 'end of stage report'. This should have space for:

- name
- candidate number
- project title
- stage title
- date of start of project
- date of start-of-stage
- date of end-of-stage
- date of report
- comment
- comment from others.

Review and evaluation

What you will learn in this chapter

You will learn how to check what you have done, ask others and test that your project meets the needs of the client. You will also learn how to develop a diary that records your work and the changes you make as you progress.

Much of your work will be in the form of a project. This chapter is concerned with how you can evaluate your work and use the results of the evaluation to better inform planning for future work.

When you have finished each stage of your work you need to check it for accuracy and relevance. Sometimes when we work on something we tend not to notice things that are glaringly obvious to others. Use others to help check you work at each stage rather than finishing and thinking that it is done. You may have gone off in a wrong direction or left the most important part out!

The whole of this book is designed to help you get through this exciting qualification. As we have put it together we have constantly checked what we are doing to make sure we match the material to the Edexcel specification.

- Always use other people's opinion.
- Constantly check your specification.
- Look for good things and try to be positive.
- Think about the examiners and how easy it is for them to find work you have done.
- Save files frequently, especially each time you make an important change.

Because you are now working in multimedia you need to make the work you have done accessible to many people with different machines and different browsers or software. You will need to share what you have done with others. To do this you need to make sure the file types you save with are correct, so they do not clog up email software.

Do your multimedia projects work? It may seem a strange question to ask but you must check several times with different users that you multimedia work actually works.

- Does your work load into a browser?
- Does your video clip play?
- Does the audio work?
- Does the animation work at the right speed?
- Does your presentation get the message across that you want it to?

Who can you use to evaluate your work

When you approach someone for advice or an opinion you make yourself very vulnerable. This is because you risk being told information that is hard to accept if your work is not of a good enough standard. For this reason, it is important that you are selective about whom you approach for such evaluative feedback. You need to be sure that whoever you ask fulfils as many of these criteria as possible:

- You respect their point of view.
- They will treat the task of giving feedback seriously.
- They will be prepared to speak honestly, not just telling you what you want to hear.
- They are fully aware of the criteria you have been set.
- They can give meaningful recommendations.
- They can be sensitive to your feelings.
- They can be diplomatic.

One way of collecting such feedback is through the use of the Anecdotal Record Form. This is useful tool which can be used to record any event that has occurred along with your own interpretation of its relevance. If you are going to use such a form, you need to make sure you complete it as soon as such an event occurs. If you fail to do so you may forget about the event or you may omit some vital description of what took place.

An expert review can be carried out. This has both formative and summative elements which can aid you to reflect on your performance so far. If you are going to ask

someone to carry out this sort of review, then you must provide them with the criteria by which your work should be judged. This is important because you do not want feedback which although well-meaning, is not actually relevant to the task you have been set.

Questionnaires are a commonly used evaluation tool. However, a poorly-designed questionnaire can cause more harm than good. So you need to take great care when constructing such a questionnaire if it is to be of any use to you in the evaluation process.

A well-designed questionnaire can be of immeasurable help to you, so make sure you pay careful attention to its structure and design.

An Evaluation Report sample is a means of presenting a summary evaluation of your work. This ensures that concise and meaningful feedback is obtained, rather than loads of irrelevant material that no one will want to read. The report should contain:

1. An attention grabbing title.
2. A description of the main elements related to the title.
3. The data relating to main elements related to the title.
4. A summative recommendation arising from the findings of the report.

When someone is given such a report they can read it easily rather than having to plough through many pages and lots of text which might require them to deduce information for themselves.

You can also obtain feedback by conducting interviews. If you are going to employ this method, then the interviews must be conducted in proper conditions that ensure the results you obtain will be sensible and meaningful. To this end it is advisable to adopt some sort of Interview Protocol so that both interviewer and interviewee know what is expected of them.

An extension of the interview method is to hold a focus group. This is where a small group of people are collected together and asked a series of questions relating to a particular product. When a new toy is designed by a toy company it is likely that it has been trialled with a focus group of children of the age the toy is targeted at. So if you held a focus group you could find out what a group of your likely users thought of what you had created. A focus group would normally be led by someone not directly involved in the project so that any findings are unbiased. In your case you might be able to ask a friend to chair a focus group on your behalf. You could then record the meeting or ask the chair to take detailed notes of the groups discussions. Alternatively you could be present at the session, but it is important that you say nothing during it.

There are many different ways of working through a project and as you get more competent you will start to take more control of the sequence yourself. However, whatever system you employ, there will always be stages. Each stage has a start, a middle and an end.

The start is characterised by setting the target, that is what you intend to achieve. For example:

	Essay	**Proofs**	**Database**
Requirement task	Write a three-page essay, of approximately 600 words, on the role played by the American media during the Vietnam War.	Print a proof copy of a paper-based publication in readiness for checking.	Add twenty records to a database.

The middle is where the tasks actually get done. For example:

	Essay	**Proofs**	**Database**
Activity	Write the essay.	Print the proofs.	Enter the records.

The end stage comprises the final tasks before you look to see what project is next. For example:

	Essay	**Proofs**	**Database**
End Tasks	Check the essay for content. Re-read the essay checkng for spelling, etc. Add formatting (if required). Take a back-up copy and print.	Package the proofs and send to the proofreader and author for checking.	Check that the data entered is correct (typed-in accurately, validated, sensible). Take a back-up copy.

Before you can start you must have a clear idea of what each stage will entail: a specification. This should be written in such a way that you can easily plan what you need to do. At the end of the stage, use the specification to check you have done everything. For example:

	Essay	Proofs	Database
Specification requirements	• The essay must be three pages long • The essay must be approximately 600 words long. • The essay must be about the American media's role in the Veitnam War.	• The proofs should be printed with double line spacing. • Print one copy for the proofreader. • Print one copy for the author.	• The records are supplied in a set format, each record contains four data fields. • Validation is automatic on two of the fields, the others must be checked manually. • Record the validity of a record in the 'Checked' field

When you are happy with the specification, you should carry out the task. When you have fulfilled the need of the specification, you should carry out a review.

The review must use the specification as its basis. For example:

	Essay	Proofs	Database
Review	• Is the essay three pages long? • Is the essay approximately 600 words long? • Have I included information on the role of the American media in the Vietnam War.	• Have I printed the proofs with double line spacing? • Have I printed two copies?	• Have I entered and checked four fields for each record? • Have I completed the 'Checked' field.

But usually it should include some aspects of forward planning: developments that may be incorporated in the future, to make the task easier, or quicker or more efficient.

	Essay	Proofs	Database
Review (possible developments)	Were any of the sources I used worth keeping for the future?	Is it possible in some cases to keep one set of proofs for my records and have the copy proofread on-screen?	Could the data entry form be better designed? Can the data be entered another way?

For a comprehensive review and evaluation, the views of others would also be recorded:

	Essay	Proofs	Database
Review (Feedback)	What mark did my teacher give me and what comments were made?	Did the proofreader make any comments?	Was the data manager happy with the data entry?

Carrying out such a review at each stage of a project is an extremely useful activity. If the project is expected to run for a long period of time, such as a term in school, it is important that regular reviews are carried out to ensure that everything is 'on track'. One of the biggest problems that traditional GCSE projects have revealed is that learners lose direction and end up doing something that is not what they set out to do.

Review and evaluation must, therefore, be built into the project plan. There may be particular stages that the project can be broken down into. These can then have 'end of stage' evaluations. There may also be time-related checks. By the end of week three, I will be able to review… .

A *Review Plan* can be a very useful tool. In business Quality Review is an essential part of any project. In many organisations there are specialists employed to oversee quality and to design quality assurance measures. One of these is the recording of reviews and evaluations in a *Quality Log*.

Other subjects already have such systems. Art students use a 'sketch book' where they record ideas, useful pictures, colour schemes or notes that they may use to influence their work in the future. Engineering students keep 'log books' where they record processes, materials or tools that they may want to use in future work.

To make your work stand out above the other learners, you could develop a quality log of your own. In your file structure, make a new folder; title it *quality log*. In the

folder, start a blog (web log – online diary) or document. Call it *DiDA Quality Log*. You should keep this file regularly updated with comments on what quality measures you have taken to ensure that your work is of the best possible standard:

This *quality log* can then be sent as part of your e-portfolio to show the examiner how important reviews and evaluations have been in directing your work.

At the end of a project you can pull together all of the reviews and evaluations easily as they will all be in one place: your *quality log*.

End of project review and evaluation

With your own *quality log* you will have a regular quality check throughout your work but you will still need to produce a formal project evaluation. This is what will be marked by your teacher and sent to Edexcel.

In your project plan, you must set aside time to carry this out effectively. Ideally you will be contacting other people and getting them to comment on your work. If that is the case, they will need time to respond, so it is no good leaving your evaluation to the last lesson!

Throughout your project you will have been making comments about the effectiveness of your work. Now is when it comes together to become a mark magnet!

Your final evaluation must show that you have carried out a **thorough review**. Linking your evaluation to your quality log will show the marker that you have carried out regular and effective reviews throughout the project, but you must also illustrate to the marker that you have tested and reviewed your final project outcome.

This means that what you produce must go through some form of testing. All of your material will have been designed with a specific audience in mind: you must test your results with the specified audience.

If you have been asked to produce an on-screen publication for pre-school children, you must test it with pre-school children. This will show the marker that you have **real** test results. Most learners will only test their work on themselves or on others in their group. Make yours stand out: test it with real people!

If, for example, you have been asked to produce a design for a poster advertising a sports centre, there are three groups you should use in your evaluation:

- The staff at the sports centre: What do they think of the poster design? Does it match with other posters or publications that they presently use?

- The customers or clients of the sports centre: Does it give them the information in a way that they find useful?
- A printer: Is it possible to produce the poster in a professional manner?

When you are asking others about your work, you should try to gather the results in a useful format. You should consider using a questionnaire or some other *formal* recording system. Try to use, 'closed questions'. These are questions that try to force the respondent to say 'yes' or 'no'. Using closed questions in your evaluation can enable you to produce statistical results.

> **Seventy five per cent of the staff at the sports centre agreed with the statement 'I think the poster matches the style of the poster we already use at the sports centre'.**

Results that generate percentages can also be shown in graph or chart form:

Another technique is to ask people to score your work. Give them a range and ask them to award a figure:

> **The sports centre customers awarded my poster 7 out of 10 for 'How easy is it to read the information on the poster?'.**

These sorts of results can be used to generate a bar chart, as each response can be recorded separately.

A third data type you could seek from your reviewers is 'anecdotal comments'. These are usually very difficult to quantify and so cannot be shown on a graph or chart, but they can be very useful to show how well your solution meets the specification. For example:

> **The manager of the printing firm said, 'I think the work you have produced here is as good as anything my staff produce. Well done'.**

This sort of comment can be added as a quote:

> **'I think the work you have produced here is as good as anything my staff produce. Well done!' The manager of the printing firm.**

When gathering your data for your review, there are a variety of methods you should employ and groups you could ask.

You may choose particular groups to harvest information from, other than the *real* users. These groups can be used to give general feedback as well as offering specific details that relate to you and the DiDA course:

- Your teacher/lecturers.
- Other members of your group.
- Your peers (other people your age).

In business, a review may also involve a focus group (a small group that is representative of the normal clients or users). You could set up a focus group such as 'Sports Centre User Group'.

Whoever you choose to use there are certain methods that can prove more successful than others.

Questionnaires were mentioned above. These are very useful for asking lots of people the same thing. If you want to get some responses from a group of customers or staff, then a printed questionnaire is an efficient method. However, if you want information from one particular member of staff, such as the receptionist or manager, a questionnaire is probably not the best way.

When seeking the opinion of an individual you should try to interview them. This does not have to be done in person, but often speaking face to face is a great advantage. Sometimes it is difficult to get together, so you could try a telephone interview.

If you are unable to meet in person, you could try to meet online (using a *messenger* application or private chat room). Failing this, you could send them your questions and ask them to respond. Always include a deadline! One major problem with not carrying out the interview in person is that the interviewee may fail to respond in time for you to use their comments.

By whatever means you carry out this type of information gathering, you must prepare. If possible you should also brief the interviewee, so that they know what is expected of them.

Before the 'meeting' you should carefully prepare your questions. Start off with some general questions to relax both yourself and the interviewee, and then ask some more specific questions. Finish by giving the interviewee an opportunity to give their general comments.

Rather than taking notes as the interview is taking place, it is sometimes advisable to record the interview and **transcribe** it later.

The whole thing should be scheduled to last no more than 15 minutes. Bear in mind that most TV interviewees with politicians or others tend to last only 5 or 10 minutes. An intensive activity such as an interview can be very hard work. Restrict what you are trying to collect to the necessary and important bits.

When you have completed the interview, be sure to thank the interviewee for their time.

If possible you should offer to send the interviewee the results from your information harvest. Sometimes this is difficult due to timing because the evaluation usually happens at the end of the project, close to the date when you will be submitting your work for assessment, but if there is time, you should send them a copy if they want it.

You should also consider sending other results to the relevant people, but rather than sending them to the individuals, you could send them to the organisation for them to display. For example, If you have been carrying out a survey of customers of a store, you could send the store the results, with an explanatory letter, for them to display to the customers or if you have asked the employees at a sports centre for their opinions concerning your poster design, you could send then a summary of the results, with a '*thank you*' letter and a short explanation of how you will be using the results in your work.

Lessons learnt

Now that you have collected information from everyone and compiled the results into charts and graphs, what do you do with it?

Most evaluations are written at the end of a project. They tend to be very superficial and might say things like: 'I really enjoyed this project' or 'I asked my mates and they said I had done a really good poster'.

Your teacher and the moderator will want to see more detail. If you have kept a record as you have worked and then carried out a survey or interview with the *real* users of your outcome, you should be able to easily put together a worthwhile and meaningful evaluation.

One part of an evaluation that is often poorly done or just missed out in GCSE projects is the 'lessons learnt' part.

There are a number of reasons for carrying out an evaluation. For example:

- To formally state whether or not the original brief has been met.
- To explain why particular aspects had to be the way they were or why changes had to be made.
- To celebrate the success of the project.
- To detail changes that might need to be made to make the solution more successful.
- To detail changes that might need to be made if the same project was to be tackled again.

The last two reasons are the ones that are often ignored. Make sure that your evaluations include them.

In industry at the end of a project the manager will often commission a *Lessons Learnt* document. This looks specifically at what went wrong and what went right, then gives a number of proposals that future project teams can follow. This is then used as the starting point of other projects.

LESSON LEARNT	
Outcome	**Suggestion**
• Poor quality of final print. • Colours are too dark and the images are not clear.	• Increase the contrast of the images before printing. • Use a test sheet before final printing. • Test print on other printers.
• Printing individual leaflets was costly and time-consuming.	• Email leaflets to reduce print run.
• Data entry took hours!	• Investigate the use of OMR technology.

Presenting your evaluation

In most cases your evaluation will be developed and presented as a word-processed document. This may then be used as the text of a web page or part of a DTP publication. However, there are other ways you may wish to try. For example:

- An on-screen publication, presenting the information in a multimedia format.
- A video, saved in an appropriate format to be viewed on a computer.
- An interactive web page, where the user can select aspects of the evaluation that they feel are interesting.

Experiment with different approaches and check with your teacher which method they think is most appropriate to the work you have been doing.

How it's done in multimedia businesses

In business, if things go wrong they usually cost the business money so they are always keen to use evaluation and review tools to minimise this. The basic plan they use is listed below and is a good guide for you too.

- **Effectiveness:** The most effective review will place the work within a broader context, explaining what important issues are worth attention.
- **Content:** When reviewing include a brief summary of the scope, purpose and content of the work and its significance in the subject. For multimedia reviews, evaluate the significance in relation to similar multimedia products and relevant literature.
- **Evaluation:** Reviews should go beyond description to evaluate the strengths and weaknesses of the work, paying attention to the use of sources, methodology, organisation and presentation. Evaluation should consider the work's stated purpose. For software on CD-ROM pay attention to user friendliness, appropriate audience, organisation and presentation. In the age of rapid technological change, the software's longevity should also be kept in mind. Also, does the use of a sofware/CD-ROM package enhance the presentation of this material?
- **Audience:** When reviewing keep in mind the audience for which you are reviewing.
- **Professionalism:** Whether the evaluation of a work is favourable or unfavourable, you should express criticism in courteous, temperate and constructive terms. When reviewing you are responsible for presenting a fair and balanced review.

Guidelines for other multimedia reviews

- Reviews should be descriptive evaluations that consider context, presentation and background. They should not be written as mere announcements.
- It is necessary to take the longevity of the material under review into account. How will this presentation or website be considered in 5–10 years? What special contribution does it make (or fail to make)?

- Website reviewers should consider not only the potential life span of the site they are reviewing, but its upkeep as well. Are links updated regularly? Is the site user-friendly? Is it frequently visited?

How to get good marks

✓ **You need to make sure you have looked through your work and thought about whether it can be improved, you need to show evidence of planning and how you managed your time. Try to be positive and include things that you felt could be improved. Use a variety of people to get feedback and help with your review.**

Homework

1. Design your own quality review folder, keep comments and adjustments in it, look to see if any trends emerge, and take note of these. For instance, you may have a tendency to make paragraphs too long or make background colours too vivid. Take note and try to avoid your more common mistakes. Keep your quality folder up-to-date.
2. Create a form in Microsoft® Word and present it as a simple review form for friends and colleagues to comment on your multimedia work. List about 10 items that are particularly important in you work and ask for feedback.

Standard ways of working

What you will learn in this chapter

You will learn how to work safely and legally and how to save your work efficiently and prevent disasters. You will also look at quality assurance and how to check grammar and spelling.

This chapter will explain why you need to work in certain ways, both when working for the DiDA qualification and generally when using ICT.

In all industries there are rules that must be followed. These rules are sometimes for safety (to protect people) and sometimes for security (to protect the business).

Many of these rules are enforced by official bodies, but many of them are very difficult to enforce. This is where guidelines are used.

Guidelines are 'recommendations'. They describe methods of working that, if followed, can help prevent accidents and make operations more efficient. But there are rarely any repercussions if they are not followed. For example, using a zebra crossing to cross a road is a sensible recommendation, but you would not get arrested if you chose to ignore the advice and cross at another point further along the road. You might get run over, but that is your fault!

Without sensible rules all systems become chaotic. It's the same in ICT.

We have all worked on a document, saved it and closed the application, only to come along sometime later and find that the file has 'disappeared'! Most people immediately blame the computer or someone else, but invariably it's your own fault. If the file had been named sensibly and filed in a suitable place, it would have been easily found later.

This chapter is all about working with ICT in an organised and logical way, following simple and sensible rules, so that not only you and your work improves, but everyone else is able to work efficiently and safely.

Before you start

Before embarking upon developing ICT-based material, you must ensure that your machine is safe and free of virus infection. A virus checker and a firewall usually protect a school machine, but as you may be sharing material between your home and school computer, you must ensure that your home machine is equally as safe.

Anti-virus software can be obtained relatively cheaply, but you may find that you can have a free version of your school's software. Ask your teacher.

Just installing anti-virus software is not good enough, you must make sure that it is up-to-date and that it stays updated automatically. This is the same for all security software.

ICT is still a relatively new field. Although computer systems date back to the Second World War, until the last twenty or so years, very few people worked with computers on a daily basis.

This has meant that ways of working in the industry are still evolving. New legislation and new procedures are developed, used and then amended, meaning that more new legislation and procedures need to be developed and so on.

The way that ICT has grown has also led to a different approach to the way legislation is drafted and put into practice. There has to be much more international agreement to make the world safer for users as well as making international trade more efficient.

ICT is often seen as a way of making things faster, more efficient and more reliable. This can only be the case if everyone works in the same way: a standard way of working.

Speed and efficiency

One obvious advantage of using ICT to carry out certain tasks is the efficency it brings. Modern processors can carry out billions of calculations per second, meaning that they can work incredibly quickly.

As you are developing your multimedia product you will come to depend upon the speed of your machine. Multimedia applications are very 'memory hungry': they need lots of space to work in!

To make sure that your applications run smoothly, you should make sure that you do not have other programs running in the background. For example, don't have a media player running while you are trying to convert video files into a particular format.

When you are using a new piece of software, try to get hold of the manual or the box. This will give a list of minimum requirements: your machine must exceed them. Your machine must have the minimum processor, memory capacity and processor speed.

New computers are being developed that have onboard sound and graphics, and even twin processing, enabling billions of calculations to be carried out every second. Unfortunately, it is unlikely that the machines you have access to in school will be at this level of specification, hence the reason for not having other applications running in the background.

To make ICT more efficient speed is only one aspect to be considered. Things such as reliability (how often a computer breaks down), whether it has connections to all the necessary peripherals, whether the software is easy to use, and so on, are also factors which influence efficiency.

The way you use ICT also has a great influence over the efficiency of the system. Using the right application to carry out a task, such as Adobe® Premiere® to edit video or Microsoft® PowerPoint® for a slide-based presentation, helps to improve efficiency.

Rules and guidelines

There are also rules that you must follow. Some are legal requirements, such as the Data Protection Act (DPA), but others are just sensible guidelines, such as taking regular breaks.

You should also consider the way in which you use the equipment and software, and the workspace that you operate within.

Workspace:

- Make sure that you have a safe environment in which to work (e.g. there are no trailing wires or other obstructions).
- Position equipment away from direct sunlight.
- Position the screen so that your neck feels comfortable (a flat screen is usually the best type and should be anti-glare).
- Use well-designed furniture to help with posture and comfort.

- The desk height should enable you to reach the keyboard and mouse easily.
- Use wrist rests on mouse mats and keyboards.
- Position equipment so you can use it without twisting.
- Use suitable lighting without glare.
- Take regular breaks from the computer (every 20 minutes or so).

If these rules are not followed, using computers can lead to:

- backache
- eye strain and headaches
- RSI (repetitive strain injury).

File management

Using a sensible method of filing is crucial. Everyone likes to work in their own way, but if people are working collaboratively (which is one of the advantages of using ICT) it is essential that everyone involved works together and uses a common approach to filing.

All ICT users must make sure that data used is accurate, consistent and reliable, and that it is stored securely.

The use of an appropriate file structure and naming convention can help to ensure that data is stored in a way that will allow easy and efficient retrieval.

All applications record the date and time of a document's use. If the user also gives it a suitable name it should make it easier to find.

The file type is also an important addition to a filename. Most operating systems have three letter extensions, that denote the type of file. This is separated from the filename by a full stop, (e.g. '.doc' for a Microsoft® Word document or '.xls' for a Microsoft® Excel® spreadsheet.

All software also generates default names, such as *document1.doc* or *sheet1.xls*, but you will have to be a little more creative when you choose a suitable name for the file.

Although the operating system can accept spaces in filenames, it is probably best to avoid them. For example, the file *birmingham bike show video.wmv* becomes *birminghambikeshowvideo.wmv*.

As this can be a little confusing, the use of 'underscore' _ can take the place of the space, so the filename becomes *birmingham_bike_show.wmv*.

As there may be more than one video from the show or you may want to collect other videos, the use of a number can be advantageous, so the filename becomes *birmingham_bike_show_01.wmv*.

Old operating systems allow only eight letters before the file extension. Although less descriptive names can be used, it can make titles easier to remember. The filename becomes *b_bs_01.wmv*.

As long as everyone knows the system, anybody who needs to have access to the first bike show video from Birmingham should be able to find it. As long as nobody types the wrong letter!

Backups

It is good practice to take backups of your data. Many systems automatically back-up data over night. Your school's MIS will probably back-up at least once a week, so that if anything happens the bulk of the data is safe.

Disaster recovery is big business. Companies that lose data through fire or theft can go bankrupt. Imagine the difficulty an international supplier would have if they lost their sales records because someone pressed the wrong button! However, if backups and other systems are put in place, a lot of disasters can be avoided.

When a system starts to get used there are only a few files, but in time this will increase to many thousands. Well over 500 files were produced to write this book!

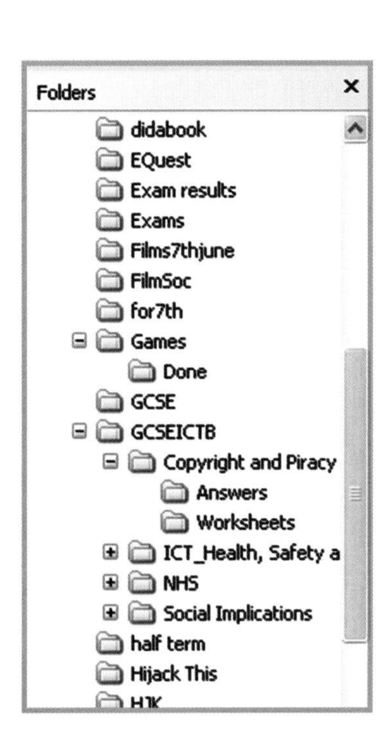

Many modern applications carry out an autosave every few minutes. This means that if anything happens, like a power cut or the battery running out on a laptop, your work is not completely lost! However this is not something to rely upon. It is good practice to save your work regularly. I have saved this document twice in the last five minutes!

Filing

Remembering filenames or the system used to generate them is one thing, but actually finding files is another. This is where the folder structure can be helpful.

Computer files can be stored in separate folders. This means that files that should be together can be grouped easily: images can be kept together in a folder named 'images'; videos can be kept together in a folder named 'videos'.

The structure containing the folders must also be carefully considered, making navigating to a particular file a straightforward operation. This skill is an essential element of developing your e-portfolio, as you need to make it as easy as possible for the marker to find each of your pieces.

It is also important that the files are stored in an appropriate format. Although Adobe® Photoshop® is often used to manipulate graphical images, the format it uses to save images is only able to be read by Adobe® applications, so if you wish to share an image it is more appropriate and sensible to save it in a format that can be read by users who do not have access to Photoshop®, such as JPEG. This may mean the image loses some of its functionality, but that may not be as important as others being able to see the image.

As mentioned in other chapters, portable document format files (PDF) are a generic file type that can be read by most machines. Saving a file in PDF can mean that it is accessible by most people. However, unless they have specialist software, they will only be able to read the file and cannot amend it in any way.

Sharing and security

If you work with others on a multimedia product, you may need to save different versions. If this is the case you can use the numbering system mentioned above: *b_bs_01.wmv*, *...02*, *...03*, etc.

It may also be advisable to set up a particular folder for shared material. If necessary the files and folder can be protected by using passwords (check the Help files of an application for guidance on how to do this for an individual file). For a folder you can use the operating system commands to share a folder over a network. Go to the properties of the folder and select the section on 'Sharing' . Depending upon your operating system, you should be able to share a folder openly with any other user or privately with particular users. There are also help sites on the Internet that can offer independent security advice.

If your colleagues are further afield, you can share materials on online storage facilities. Again, these can be secured with passwords.

Bear in mind that any files that contain personal data fall under the Data Protection Act and you must therefore ensure that they are kept securely and that access is limited to individuals that have been given authorisation.

Remember that any system is only as secure as the people using it want it to be. A password is a great way of stopping people accessing materials they are not supposed to, but if the password is not kept secure the files won't be!

This is similar to credit card Personal Identification Number (PIN) security. PINs are secure only if the user does not give their number to anyone else or leave it for others to find.

If a folder has been set up to contain a number of specific files it is sensible to list the files and their contents, and give instructions on what to do with them and who to contact about difficulties. It is good practice to do this using a 'readme' text file. A text file usually has a '.txt' format, so that it can be opened in a range of applications.

Software effectiveness

If you can remember when you first started using ICT, you may remember that everything took a lot longer. Machines have not got that much better – you have! You will have learnt some shortcuts such as using 'Ctrl+C' for copy and 'Ctrl+V' for paste, or using 'F1' for accessing the application Help files.

Every software package has a vast array of helpful shortcuts, these are easily found when you look at a menu item. The menu bar itself has an underline character in each title and the menu items also have one letter underlined. By pressing the 'Alt' key and the underlined letter in the title, the menu is expanded. Then simply press the letter on the keyboard corresponding to the underlined letter of the menu when you want to perform. For example, to show the 'Task Pane' in a Microspft® application: Press ALT+V followed by K.

Some functions are common across applications, such as copy and paste, but some only work in particular pieces of software.

Applications such as Adobe® Premiere® not only have shortcuts on the menus, but the developers have made specialist keyboards with coloured keys and other labelling. These keyboards can prove very useful if you intend to use such applications regularly.

'Shortcuts' can also be used to access common functions. If you look at your desktop you will see a number of shortcuts which, when clicked, carry out a particular function, such as opening a folder or starting an application. They are very useful tools for speeding up how you use a computer. They can be created by right clicking on an item and choosing 'create shortcut'. The small file this creates can be cut and pasted to another location, but it will remember what it is a shortcut to!

Adding shortcuts within a filing system can also be useful because the shortcut can connect the user to files which are stored elsewhere on the system without having to move or duplicate the files.

It also helps if you can choose the correct tool for a job. In manufacturing a skilled worker is one that is most able at selecting and using the appropriate tool, correctly. This is the same in ICT. Selecting the correct application and tools within an application will make you a much more efficient user of ICT.

> PRACTICE MAKES PERFECT:
>
> You may have come across different computer mouse configurations; one button, three button, buttons and wheels, roller balls, joysticks and all manner of other things to enable you to control movement on the screen. These each have their role, some are flexible and can be used in a general capacity, others are designed to be good at a particular task or for controlling a particular application.

Quality frameworks

A lot of good practice is common sense. Most of what has been mentioned above is sensible good practice. But to ensure that standards are raised it is important to work within a quality framework.

> Don't just copy chunks of data from encyclopaedia sites. The skill is in being able to adapt the material to your needs. Copying and plagiarism are easy to detect.

With modern DTP applications there is no excuse for poor spelling. Each software package has a spell checker and most will also check grammar. Some systems have thesaurus and dictionary functions or even links to encyclopaedia entries. But even if the application you are working with does not have all this functionality, if you have access to the Internet you can get help with any aspect of your work by accessing websites that can perform checks for you, such as www.dictionary.com. But be careful! Some dictionaries may be American and give the US spellings for English words. For a similar reason, make sure the language setting is for UK English.

Checking

However, even with all of the help tools working and checks being run constantly, you still need to check material for yourself. It is essential to reread all documents, not just text documents. You need to check that audio and video illustrate the right points, slide transitions all work when expected, and so on.

If a document is to be printed you must also make sure that before printing you do a 'print preview'. This should be done every time a file is updated and prior to printing. Remember trees die for all the paper you use!

Chapter 9 mentioned the need for getting other people to check your work, such as a sample audience. This can be a very useful tool because you may not spot what other people find obvious. You may have spotted an error in this book that we and our checkers never noticed. Sometimes you can be 'too close' to your work and the obvious goes unnoticed.

You must also make sure that you recognise when others have helped you and record that in your work. This also goes for acknowledging any sources of information and authenticating your own work. You will be submitting an e-portfolio for assessment that will have to be authenticated, proving that the work within it is yours. However you must also point out where and how you have been helped.

By authenticating your work, you are stating that you have worked within standard ways of working and, in doing so, have complied with all appropriate rules and regulations. You have acknowledged where you have used other people's work or ideas and sought and received help. You have also ensured that material that is confidential has been kept secure and only those with a particular reason have had access to it.

As mentioned above there are certain laws and guides that you also need to work within. The Data Protection Act is most often mentioned where ICT is concerned, but there are other laws that you need to be aware of.

- **The Data Protection Act (DPA):** Controls how personal data is handled and distributed. It is essential reading for anyone dealing with personal data.
- **The Health and Safety at Work Act:** Covers health and safety in the workplace. This ensures that accidents are minimised. Just because you don't work there doesn't mean it's not a workplace! Other people do.
- **The Disability Discrimination Act (DDA):** Related to access for all, ensuring that people are not disadvantaged through physical disability or learning difficulty.
- **The Special Educational Needs Discrimination Act (SENDA):** Linked to the DDA, but more specifically for those with learning difficulties.
- **Copyright and Patents legislation:** Covers all published materials and manufactured goods. If something is used for gain, it must be credited, and in most cases paid for.

How to get good marks

✓ You need to show the examiner that you have used clear and sensible filenames, used backups, chosen the correct software and techniques, acknowledged sources and respected copyright. You also need to show that you have worked safely.

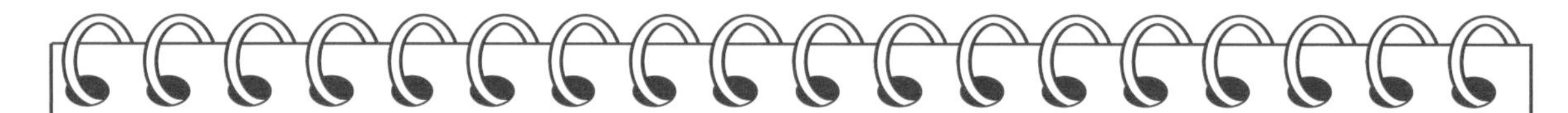

Homework

1 Carry out an audit of your workspace. What can be done to improve it? Produce a multimedia page highlighting some of the dangers faced by computer users, such as poor seating position, equipment positioning and cluttered desk space.

2 Use the Internet to find out about the laws covering copyright of audio and video. Produce a multimedia page explaining the laws, such as what can be done legally, what you should avoid doing, what must not be done, and who is responsible.

3 Produce a set of rules that could be used as the basis of a contract between you and your teacher to ensure that you worked safely and efficiently.

Artwork and image-editing software

What you will learn in this chapter

You will learn about image-editing software, scanning, artwork, names of tools common to this type of software and how to create exciting text and images.

Creating multimedia products will almost certainly require you to use images. The images may be used on a website, for placing onto CD-ROM, to create software or in a presentation. Image-editing software is specifically designed to allow you to change file types (Adobe® Photoshop® is particularly good at this), at changing image sizes, and at changing lighting, colour and backgrounds, plus almost anything else you could think of doing to your image. Image-editing software tends to be very powerful and allows you to be very creative with images.

What is artwork?

Artwork in the printing and graphics industry refers to any work that is finished and is ready to be published or printed. It can be text on its own or text and images or images on their own. In an image-editing package you can put the final touches to an image before publishing.

Image editing

No matter which means of technology is used, digital images can be captured with varying degrees of quality. It is often the case that images captured digitally will have to be edited using graphics software.

It may be the case that an image is too dark or light, or that a person has 'red eye' all of which might render an otherwise good image useless. The joy of graphics software

is that stunning remedial effects can be achieved with relatively little effort. Such software can make life so easy for a designer that it has to be a vital component of such a person's software library.

Some specific goals of image editing are:

- to crop, rotate or otherwise manipulate the orientation of the image;
- to remove or mask imperfections that might otherwise spoil the effectiveness of the image;
- to compensate for incorrect levels of brightness, contrast and other colour imperfections;
- to improve the balance of the image;
- to sharpen the image.

What is a scanner?

Flatbed scanners are indispensable pieces of equipment that allow the user to convert any image into digital form. These devices are now extremely cheap and can scan in a whole range of resolutions and colour ranges depending on the resolution required. Images that are captured in such a way can then be imported into graphics software where they can be altered as required.

Optical Character Recognition (OCR) software can also be used in combination with a scanner, in order to capture text, which can be imported directly into a word processor.

Scanning software packages have different features and some allow more manipulation than others. It is sometimes best just to capture the image using scanning software and then carry out further manipulation within specialist graphics software packages.

Image-editing software can change the colours, move selected parts of an image or add effects. In fact the more powerful image-editing software, such as Adobe® Photoshop® allows you to do just about anything to an image you can imagine! However they are not just able to change images. Image-editing software can produce amazing typography (letters). You can even paint or draw and use the effects option to create brilliant designs. Most image-editing software is used to prepare images ready to be printed or to be included in to web pages. It is part of your creative toolbox.

Whether you take photographs with a digital camera, scan pictures or create your own original artwork, you'll need a good imaging tool to tweak your pictures. Low-end packages let you do basic visual manipulations such as adjusting brightness or cropping the picture. Microsoft® Word allows you to scan an image then adjust it using the image control tools on the highlighted work. You can also crop poorly composed shots using the cropping tool by dragging it over your work until it is the right size (see the example on page xx).

In industry standard software packages such as Adobe® Photoshop® you can do much more! Adobe® is the software publisher best known for graphics. Their imaging software includes Photoshop®, ImageReady® and Illustrator®. Other companies such as Macromedia® also produce popular software such as Fireworks®. Lots of other tools can be used for imaging including CorelDRAW. Some packages, such as AppleWorks®, contain paint and draw options within their software.

Capturing images

You can capture images in many ways

- with scanners;
- with cameras;
- by copying from the Internet;
- by buying images from image libraries (expensive);
- from clip art libraries.

There are two types of scanner you are likely to come across: the flat bed and the feed.

With a flat bed scanner you lay a book or image face down on a glass plate beneath which a bright light source and detector is moved. This type of scanner is ideal for thick or larger images.

The feed scanner in usually incorporated with a printer. The image is fed through the scanner which has a stationary light source and detector. You can only scan images that are on thin paper or card and that are no larger than the paper size the printer can take (usually A4).

Most scanning software allows you to do a pre-scan. A pre-scan scans the whole bed then

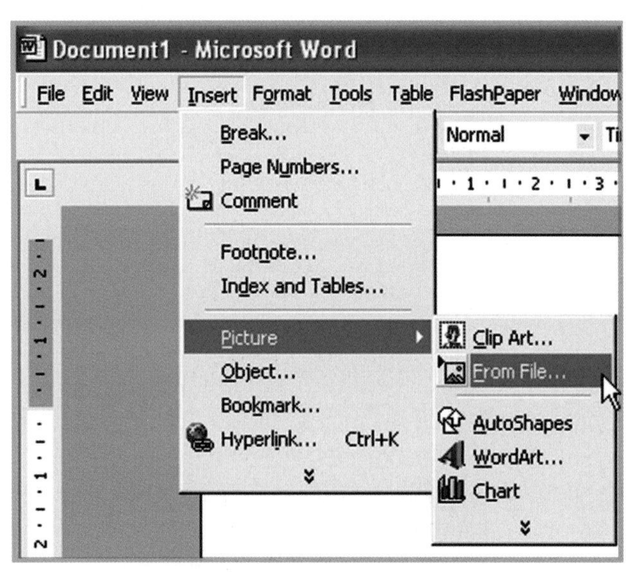

allows you to make a decision on size or cropping. The resolution can also be set (300 dpi (dots per inch) is usually fine, but for Internet images it can be less). The higher the resolution the more storage space it will take up on your computer. You can set your scanner to work directly from many applications.

Adobe Photoshop®

There are many specialist graphics manipulation packages, but Adobe Photoshop® is perhaps the best-known and most popular. A scaled-down version of this package, called Adobe Photoshop Elements® is also available. The package can be used to achieve great effects with files saved in a range of different formats.

Adobe Photoshop® Basics

In common with most graphics packages, Adobe Photoshop® has tools for elementary manipulation, such as cropping, rotating and resizing.

Whether using a PC or Apple computer, a range of keyboard shortcuts enable the user to carry out an array of effects with ease.

Some of the most common shortcuts are (Apple users should substitute the 'command' key for the 'control' key):

Ctrl + A	Select All
Ctrl + C	Copy
Ctrl + V	Paste
Ctrl + S	Save
Ctrl + D	Deselect
Ctrl + Z	Undo
Ctrl + U	Adjust Hue
Ctrl + X	Cut

As you can see some are more obvious and memorable than others.

The larger the image, the slower your computer will run, so it makes sense to keep images as small as possible, without compromising the quality of the image. There are many methods for resizing images. Here is one suggestion:

How to resize an image

Lets take an image 584 pixels high by 400 pixels wide which we need to resize to 200 pixels high by 300 pixels wide.

1 Select Image > Image Size from the Adobe Photoshop toolbar. The resulting dialogue box will show you the current width, height and resolution of the image. At this point it is important that you ensure the Constrain proportions checkbox is ticked. This will keep the image in proportion no matter how you manipulate it. If you fail to tick this box the image will be distorted and it can be hard to return to the original image dimensions without starting all over again. Use Ctrl+Z to restore the original dimensions if such a problem arises. Then select Image > Image Size again.

Another advantage when the Constrain Proportions box is ticked is that you only need to enter one of the dimensions you wish to change, as the other will alter automatically. In this case the dimension for height needs to be altered to 200.

Now the software will automatically alter the width to 137 pixels.

2 Now to change to width of the image. This is best done by choosing Image > Canvas Size which adds pixels to the image by filling them with whatever background colour you choose, whilst leaving the image unaltered.

In this case the width is altered from 137 pixels to 200 pixels.

The Placement display lets you see where the image (the darker box) will be in proportion to the background (the lighter boxes). By default it sits in the centre, which is where we will leave it.

How to create text effects using filters

1. Load up a new file, 250 pixels high by 50 pixels wide. Using the Type tool, and a bold font, type some text. This is created as a new layer which you can then alter using filters. To begin this process you need to choose Layers > Render Layer from the menu.
2. In Adobe Photoshop® 3 you can apply the Alien Skin Cut-out filter. In Adobe Photoshop® 4 and 5 you do not need to select the text, just apply the Cut-out filter.
3. Choose the background layer and make the background colour white, the foreground layer light blue. Make use of the gradient tool to make a graduated background. Using the shift key the gradient will be constrained to a 90-degree angle.

How to use custom brushes

Open a new file with the Brush Palette open, double-clicking the brush icon if it is not open. This palette allows you set a brush size and it is advisable that whatever size you choose, you should ensure that it is still manageable.

In order to make a brush pattern of your own, open a file 100 pixels x 100 pixels and choose Filter > Add noise, using monochrome setting and use Ctrl+I to repeat the noise filter.

Soft edged brushes tend to be of more use than brushes with hard, defined edges. To achieve this effect, choose an area and then Select > Feather, Select > Inverse, then fill the area white.

Now decide what area of the image you want to use as a brush. Then select Define Brush from the brush palette. This will create a new icon for the brush you have just created.

Brushes created in this way can be saved by selecting Save Brushes from the palette.

You will find it easier to add to the pre-existing brush palette rather than trying to create a brand new one of your own. You should experiment with your brush to see that you get the desired effect and if it does not work as expected, you can always delete it from the palette by selecting the rush and choosing Delete Brush.

Threshold mode
This is used to pinpoint the lightest and darkest portions of an image.

To activate threshold node open the Curves dialogue box, dragging the pointer to identify the darkest and lightest points.

Open the Levels/Curves dialogue box and double-click the pipette tool to show the colour picker. Then choose a value for the area you identified.

About grouping and ungrouping objects

Grouping is an extremely useful feature which allows you to carry out a specific action on several objects simultaneously. This can be achieved by one of two methods:

- Holding the shift key as you select objects.
- Dragging a box/marquee over the items you want to group.

To work with grouped objects you can use the tools in the Drawing toolbar which is found in the bottom left hand portion of the screen. To display the toolbar choose Customize on the Tools menu, then click the Toolbars tab. Click Draw then Group.

To ungroup objects, select the grouped object and click Ungroup. The same process can be used to regroup objects, but this time you select Regroup from the same menu.

Using the clipboard

You can put text onto the clipboard by highlighting the text and using Ctrl+X to cut it or Ctrl+C to copy it. The text can then be pasted using Ctrl+V.

There are many ways in which text can be manipulated.

How to flip a shape

Select a shape then either point to Rotate or Flip and choose the orientation you require.

How to rotate a page

Rotating the page allows you to obtain interesting and unusual effects such as having objects that are not at the traditional viewpoint, but appear to be three-dimensional. To rotate the page press Ctrl and press the mouse on one corner of the page. As long as the mouse button is depressed, the pointer will change appearance and become a rotational pointer. You can then physically move the angle of the page as you manipulate the mouse.

It should be noted that the rotational effects do not alter the way that a page is printed.

How to get good marks

✓ **You need to show the examiner that you know and understand the value of capturing and using images from a scanner and a digital camera, that you know how to manipulate and modify images and text and that you know the different needs for on-screen and printed publications.**

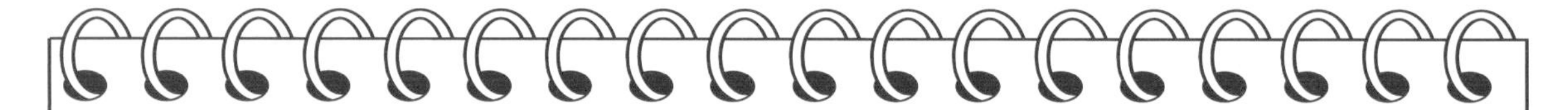

Homework

1 Go to Picasa. Picasa is free image editing software that can be downloaded in Google. (Log on to the Internet go to Google and download Picasa free)?

Try some images you have saved by placing them into Picasa. Have a go at editing them

Picasa is software that helps you edit and share all the pictures on your PC. Every time you open Picasa, it automatically locates all your pictures (even ones you forgot you had) and sorts them into visual albums organized by date with folder names you will recognize. You can drag and drop to arrange your albums and make labels to create new groups. Picasa makes sure your pictures are always organised.

.http://picasa.google.com/index.html you can download the software for free. It is there for you to start image editing and creating wild effects.

Digital video and audio

What you will learn in this chapter

You will see how video and sound can be used in the DiDA work and why you should use it. You will find guidelines to help you design an effective presentation.

Putting video on your website or presentation

The difference between an entertaining video and one that no one wants to see is usually in the care and skill that went into the creation of the video. If you have a large quantity of video, take a few moments to plan and edit it to create something really special that appeals to you and others. Be warned working with video is nearly always going to take you a long time. The quickest way to make sure you do not waste time is to plan out what you are going to do. This will save you time in the long run. Know what you want and understand the needs of your audience and try to work *for* them.

There are a number of ways of getting video into your computer. You can download movies or you can transfer data directly from a video camera. Once the material is in your computer you can soon start editing images.

The most important thing to remember about video is that it can take up a lot of disk space. Saving onto CD once you have finished downloading may prevent your computer from slow running or even crashing.

Multimedia video formats

Video can be stored in many different formats.

The AVI format

The Audio Video Interleave (AVI) format was developed by Microsoft®.
It is supported by all computers running Microsoft® Windows® and by all the most popular Web browsers. It is a very common format on the Internet, but it is not always possible to play AVI material on Windows® computers without Microsoft® Windows®. Videos stored in the AVI format have the file extension .avi.

The MPEG format

The Moving Pictures Expert Group (MPEG) format is the most popular format on the Internet. It is supported by all the most popular Web browsers and can be used on a Mac or a PC. Videos stored in the MPEG format have the extension .mpg or .mpeg.

The QuickTime™ format

The QuickTime™ format was developed by Apple®. QuickTime™ is a common format on the Internet, but QuickTime™ movies cannot be played on a Microsoft® Windows® computer without an extra component installed. This component can be downloaded free. Videos stored in the QuickTime™ format have the extension .mov.

The RealVideo® format

The RealVideo® format was developed by RealNetworks®. The format allows you to stream video (online video, Internet TV) with low bandwidths. Because of the low bandwidth, quality is often reduced. Videos stored in the RealVideo® format have the extension .rm or .ram.

The Shockwave® (Flash®) format

The Shockwave® format was developed by Macromedia®. It requires an extra component to play material. This component comes already installed with the latest versions of Netscape® and Microsoft® Internet Explorer. Videos stored in the Shockwave format have the extension .swf.

Audio

Microsoft® Windows® Sound Recorder is a simple audio recording utility included with all versions of the Microsoft® Windows® operating system published since 1995. It is described in great detail in Chapter 2.

Although it may seem like a good idea to have an audio track to accompany your project, it is possible that this might alienate those who do not speak the same language as you or people with hearing impairment. As a result the audio track has to be clear and unambiguous. It is advisable that the speaker has good diction and that the audio is recorded using the highest quality equipment available. An alternative to this is the provision of a subtitles or text track, but small computer screens can render such tracks impractical.

The basics

Nothing attracts a viewer's attention like a good quality video clip. However, as video can take a long time to download to a computer it should be used sparingly on any website. It is also important that the quality of the video is optimised to take account of the speed of the user's internet connection. The most crucial question you can ask is: is this clip really important? If the answer to the question is 'no', then you should not use a video clip in that circumstance.

If someone spends the time (several minutes) it will take for a video clip to download, they will expect it to be worthwhile. Therefore, it is advisable if you give them some idea of what to expect by either summarising the content in a paragraph of text, or by providing some still images from the video. If possible you should cut long videos into smaller, more manageable chunks so that they download more quickly.

Editing video

A wide range of software of varying degrees of complexity exists with which video footage can be edited, including:

- Adobe Premiere
- Apple iMovie
- Adobe After Effects
- Microsoft MovieMaker

Adobe Premiere® is a dual platform package which yields professional results. In particular this can be used to import several video clips and edit them together. This package can also be used to include still images created in packages like Adobe Photoshop®. A series of filters and transitions then combine to make the video an even more attractive proposition. Premiere includes features that make is possible to synchronise audio with your video footage.

Advice on sound

An aspect of presentations that is often overlooked is accompanying sound. Quite often a simple sound effect at the right time can add an important degree of interest on the part of the audience. However, it is important to remember that not all users may have immediate access to sound unless they have headphones, so you must not rely on audio tracks to convey important content, unless you warn your audience of this fact beforehand.

Uploading video

It is a rather straightforward, if time-consuming process to upload video to the Internet. Prior to doing so, it is vital that the video is optimised by ensuring it is in a format and resolution that is suitable for the Internet. This will mean that the video quality will be reduced, if this is not done the video file will be so big that it will take too long to download.

You should view your video over a range of different Internet connections to ensure that it performs as desired on all computers.

Some video guidelines

- Take care not to present a video just for the sake of it. For example, voice output only can be as effective as, and requires less storage space, than a video of someone speaking.
- Using video as part of a multimedia application usually requires quality as high as that of television to fulfil the expectations of your users.
- Use of techniques such as cut, fade, dissolve, wipe, overlap, multiple exposure, etc. should be limited to avoid distracting the user from the content. (In other words, don't get carried away!)
- Use short sequences. This is different from watching a film, which usually involves watching it from beginning to end in a single sequence. Video sequences should be no longer than 45 seconds. Longer video sequences can lessen the user's concentration.
- Videos should only be accompanied by a soundtrack in order to give extra information or to add specific detail to the information.
- Videos need time and care if they are to present information clearly and attractively.
- Lighting conditions under which a video is to be viewed may be poor, so controls should be provided for the user to alter display characteristics such as brightness, contrast and colour strength.

- Provide low quality video only within a small window. Full screen video raises the expectation of the user. Often some kind of 'decoration', for example a cinema or theatre curtain can be used to show low resolution video in a part of a screen.
- The point the viewer has reached in the video or animation sequence and the total length of the sequence should be shown on a time scale.
- The user should be able to stop the video (or animation) at any time and to rewind/fast forward it. The most important controls to provide are play, pause and replay from start. However a minimum requirement is that users should be able to close or cancel the video or animation sequence at any time.
- Video controls should be based on the controls on a VCR or hi-fi as these are familiar to most people.
- You should provide controls to set video characteristics such as brightness, contrast, colour and hue.

How to get good marks

✓ **You need to show that you know how and when to use sound and how to use video clips that you have captured.**

Homework

1. Run Microsoft® Movie Maker if you have Microsoft® Windows® XP and try out some sample videos. Alter the colour, brightness and any other effects then replay.
2. If you have microphone run Microsoft® Windows® Sound Recorder and record some simple tracks over tracks to build up a simple song or story line.
3. Download a few movies from the Web. Look at the techniques used and take notes. Most you will find are simple and straightforward without too many effects. Some will use effects. Why is this? Make some notes and compare with others.

The Internet and intranets

What you will learn in this chapter

You will learn about the features of browser software, bookmarks and favorites, how to use a search engine and how to work efficiently.

This chapter is about how the Internet works and about viewing web pages efficiently.

The Internet

The Internet is a huge network of computers. In fact, it is a huge network of networks. Every computer that logs on to the Internet does so through some sort of provider. This gateway is the link from one network to all the others.

The concept and forerunner of the Internet arose in the 1960s to allow research laboratories and universities to work collaboratively.

In developing multimedia products, you will need to consider how your material will be best viewed by your client. It is likely that much of your work will need to be shown on the Internet or on an intranet.

Intranets

Intranet systems are basically 'closed' networks that run exactly as the Internet does. The only difference is that they are not open to the public.

Sometimes an intranet is confined to a geographical area, such as a town, but often it is kept within a company premises. There are some intranets that can be accessed by anyone, but they usually require a password.

Intranets are really useful for private and confidential material that needs to be transferred over a network or if use needs to be monitored, such as in a school. Most schools and most large organisations have intranets. Some have developed extranets (intranets that can be accessed remotely through a standard Internet connection).

The World Wide Web

The most commonly used part of the Internet is the World Wide Web. This is the system that can deliver images, text, audio and video from anywhere in the world to any suitable computer.

There are other systems that transmit data over the Internet, that are not part of the World Wide Web (e.g. email).

In order to have access to this massive resource, computers have to have a common language. The language most often used is Hypertext Mark-up Language (HTML). This is an electronic language that computers use to display web pages. Hypertext transfer protocol (HTTP) is used to communicate across the Internet.

Other programming languages, such as JavScript can display web page content. They are often embedded in the HTML.

To view the programming language in which a web page has been written, click on View > Source in Microsoft® Internet Explorer.

Multimedia products usually involve higher levels of HTML or other languages than purely text-based pages, and they tend to be quite complicated. For DiDA you are not expected to be able to program in any particular language, but if you want to work in the multimedia industry when you complete your studies, you should try to find out more about 'coding'.

There are many websites and books offering explanations of what the coding statements mean. The software application itself will often have a reference section to help you. Alternatively, view the source code of a few websites and see if you can work out what some of the more common statements mean.

> Look out for < and > symbols and the comments that are entered next to them.

Browsers

For a computer to be able to understand any of these languages it must have an application installed that can code and decode the data. This is called a **Browser**.

The most common browser is Microsoft® Internet Explorer. Other PC browsers are Mozilla™ Firebox®, Opera and Netscape®. Apple® Safari™, Microsoft® Internet Explorer and Mozilla™ Firebox are all popular Mac browsers.

Whichever system is used the principle is the same. An analogue or digital electronic data stream is decided by the computer and the resulting code is passed to the browser application to be rendered as the web page you view. If the source data is analogue you will need to convert it, using a modem, to digital serial data so the computer can use it.

A modem (MODulator DEModulator) is a piece of hardware that converts analogue signals to digital and vice versa.

Modern browsers are able to handle a wide variety of file types and media by using **plug-ins**. These are small programs that are added to the browser application. The browser uses the plug-in to translate the code of a particular file type.

As you develop your multimedia product, you should consider whether a standard browser will be able to show your work and, if it can, does it match with what you want users to see?

Unfortunately, many multimedia products need a variety of plug-ins before a browser will show the page. If your materials need plug-ins, you should consider including links on an standard HTML page so that the user can get the appropriate software quickly.

For example, if you have developed a SWF file, from a Flash® project, anyone wishing to view your work will need an up-to-date Shockwave® Player. This should load automatically, but some browsers may have their automatic update function switched off. For these users a direct link to the Macromedia® Shockwave® page is essential.

Addresses

To be able to access information over the Internet you need to know where it is!

If you walk into a library and simply ask the librarian for a book you will probably be laughed at! If you ask for a book about dinosaurs, you will be directed to a range of books, but it would be even better if you ask for a particular book, by a particular author. The librarian can then provide exactly what you require.

If you want a precise piece of information you need to be able to ask a direct and accurate question: 'please can you tell me, what is the third word, on the fifth line, of the last page of Romeo and Juliet?'

The Internet is just like a massive library. It has pages covering every subject you can imagine, from aardvarks to zebras and everything in between!

To make sure you have a fruitful time on the Internet you must be prepared. Make sure you know what you want or where you can find it.

Websites all have addresses: Uniform Resource Locators (URL). These are like household addresses and they are made up from particular strings of data. If you can understand how they are made up, you stand a better chance of using them efficiently.

Take the address http://www.hoddereducation.co.uk/faq.aspx

This is made from a number of elements:

http:// stands for hypertext transfer protocol. It is the system that delivers data over the World Wide Web. The prefix lets the browser know that it will be receiving hypertext pages. Other prefixes indicate different protocols.

www. This indicates that the host server (where the website is located) is running on the Internet and are public files (not intranet files).

hoddereducation.co.uk is the 'domain name'. It is usually a name that has been purchased and registered by someone, or a company or organisation. This is the name of the top level of the file structure where the web server will look for the requested pages. The identifier .co.uk is available for companies based in the UK. There are lots of others. For example, .com is the most common as it relates to companies that can be based anywhere, .gov.uk is the extension used by sites that are Government run .ac.uk are colleges, and .sch.uk are schools.

faq.asox is a subdirectory (or page below the top-level home page).

The URL gives us an 'easy to remember' way of finding websites. The websites are uniquely identified by an IP address.

Browser controls

As with all Windows® applications a browser has a Graphical User Interface (GUI) that is designed to make most operations straightforward, usually with only a few clicks of the mouse. The GUIs of many browsers are similar.

A title bar at the top of the browser gives the title of the page you are viewing and contains the minimise, maximise and close buttons.

The menu bar contains File, Edit, View, Tools and Help links. There is also a menu item that allows you to quickly go to 'often used' sites.

Although different browsers are available they all do similar jobs (display Web pages), therefore, their menu and toolbars are similar. However, some customising is possible.

Here are some toolbar functions you will find on most browsers:

- **Back:** Return to the page you viewed immediately prior to the one showing in the browser.
- **Forward:** Go forward to the page you have just gone 'back' from.
- **Stop:** Stop whatever is happening. This is useful if you have clicked something by mistake.
- **Refresh:** Pages can load incorrectly or information may need to be updated. This gets the latest version of the page.
- **Home:** This is the page that is shown when the browser starts. It can be set to any URL. On an intranet system this is often a log in page.
- **Search:** The most used feature of the Web is its ability to find information .
- **Bookmarks/Favorites:** This gives you access to the URLs you have stored for quick access.
- **History:** This records every URL that the browser displays. It is useful if you find something and then need to go back to the source at a later time or date. Remember the history log has a stored time limit!
- **Email:** Email and the Web both use the Internet's infrastructure. It is possible to send or receive emails whilst browsing.
- **Print:** It is sometimes useful and necessary to print a web page, but bear in mind that web pages are not all the same size so always preview before printing.

Dynamic HTML and other web-based languages allows pages to be designed that are 'changeable', interactive/and customisable.

A printed book page is 'flat'. It can be read, but that is the limit of its functionality. A web page can look like a page from a book, but it can do so much more.

A web page can contain a variety of different media:

- **Text:** Typed text, formatted as it would be in a word-processed document.
- **Images:** Photographic quality pictures, manipulated to be appropriate for viewing on the screen.
- **Vector images:** Shapes and colours used to enhance the appearance of the page.
- **Animated video:** Clips using vector graphics.

- **Full motion video.**
- **Audio.**
- **Games.**

And just about anything else that can be transmitted electronically.

But even this is only like watching a television as opposed to reading a book. What makes websites potentially more interesting is the ability to interact with what is going on using interactive components such as forms, buttons and hyperlinks.

Any part of a web page can be turned into a hyperlink. A hyperlink is often associated with text, but it can be anything: text, images, animation or a combination.

As you develop your multimedia product you need to consider what hyperlinks you will be using and how you can make them obvious to your audience.

Standard web pages change the mouse pointer as it moves over a link. More advanced pages change the pointer in different ways, such as changing the colour, adding tails, even leaving a trail of bubbles or snowflakes! Although you could spend ages trying to develop these things for your material, it is a lot quicker to search for free code on the Internet. Many developers are happy for others to use their little tricks, as long as they get an acknowledgement.

Your multimedia product will be expected to include elements such as 'rollover buttons' or 'disjointed rollovers'. You will also need to develop a working navigation system, this may include a variety of techniques, but check that they will work in a browser window before you go too far.

Data capture

A web page's ability to capture movement and action can be exploited with forms.

A form can be built into a web page for a user to complete. The data is then captured by the browser and can be sent to the web server. In this way web-based material can become truly interactive. The web server can automatically search for a subsequent page based on the user's input and deliver information back to the user. This can let every user have a different experience on the same website.

This ability also allows web pages to initiate searches.

Searching

Many websites contain complex search systems. These are often built using specialist programming languages that operate within HTML pages. Some of the most common languages used are Active Server Pages (ASP), Structured Query Language (SQL), and Macromedia® ColdFusion®.

It is unlikely that you will need to add this functionality to your multimedia product, but as with languages mentioned earlier, if you want to work in this field in the future it might be worth trying to find out more. There are some very good tutorials on the Internet and contained within the Help files of some applications.

If you are using Microsoft® FrontPage®, there is a facility for adding a search to your own websites. There are also a variety of other functions that can be added using this application. However, be careful not to add everything you can find. It can make your work look messy.

Search engines, such as Google® or Ixquick, work by the user entering a search criteria. The web browser sends this request to a web server application. The web server application then carries out the search using specialist software.

As with searching a database, or running a query, the more detail you can add to the search criteria, the more likely it is that you will get a good result.

As mentioned earlier, simply asking for a book will not be a very good request, so when searching for something specific, use specific criteria.

If wanting to find the price of a new Ford car in Essex, a search should include a number of words, perhaps: Ford new Essex

Capitalisation is irrelevant, but punctuation does make a difference. A full stop means 'and' so if you want the search for the three words use full stops between Ford and new, and new and Essex. Making the search engine look for a web page containing 'Ford and new and Essex'. It will not return a page that has only 'Ford and Essex'.

There are other tricks to helping organise your searching:
+ in front of a word means 'is required'.
– in front means 'is not required' (meaning it must not appear in the result).
Quotation marks will restrict the search to finding exactly what appears inside the quotes.
* (asterisk) can be used as a 'wildcard' standing in for letters or words when you are unsure of what might be in the search criteria. d*ve will make the search look for words that have d v e and the second letter can be any number of letters of numbers or characters.

There are different search engines, some of which specialise in particular areas of interest. The most popular general search engine is Google® (www.google.co.uk), but AskJeeves! (www.ask.co.uk) and ixquick (www.ixquick.com) are also popular.

There are also online telephone directories, (e.g. www.192.com), business directories (e.g. www.scoot.co.uk) and all manner of others. Try entering 'specialist search engines' into a search engine mentioned earlier!

Saving and printing

When you have found the page you want by searching or simply by browsing, and it shows information that you think is useful, you will want to do something with it.

If you think you may need to come back to it at a later date, rather than using the 'history' function (which may 'time out'), you could save the URL in your bookmarks or favorites list. This stores the URL you can click on it to open the page at a later time.

You can also save a page 'for offline use'. If you do this, a copy of the page is taken and stored as a temporary file so that you can access it through your favorites/bookmarks list, but it will not be updated if the 'real' page changes.

You may want just part of a page. In this case you can copy the particular section, as you would with any other document and paste it into a word-processed document. However this will normally only copy material associated with the standard HTML of the page. Anything like video or sound may be lost, as will much of the interactivity.

Copying and pasting is probably the most useful way to collect information from a website for use in your own work, as you can quickly gather small 'chunks' of useful data. But you must remember to record where you got it from, so that you can acknowledge it in your own material.

Video and audio are always a bit tricky, but if you contact the webmaster or the company that runs the site, they may allow you access to the files or even email you a link to where you can download the files.

If you print a web page, preview it first. If it seems that something is not going to print, it may be possible to take a screenshot and then paste it into a document.

How to get good marks

✓ **You need to show that you know how to use the features of Internet software, that you can use search engines and wildcards efficiently, and that you can create hyperlinks within websites.**

Homework

1. Collect the addresses of a range of websites that offer free downloads of useful code, such as trails and other mouse pointer effects.
2. Develop a welcome animation that plays when a user clicks a hyperlink. It should be a small file so that it does not take long to download. It should also work without any specialist plug-in. Be sensitive to the fact that animation may look unprofessional and deter clients!
3. Collect the web addresses for sites that offer plug-ins or codecs that you may need to offer to your audience when you distribute your multimedia product.

Presentation software

What will I learn in this chapter?

You will learn how to build a PowerPoint presentation and how to avoid producing a poor slide show. Detailed skill sections show you the basics of PowerPoint and how to relate the slide show to the target audience.

Presentation software such as Microsoft® PowerPoint® is truly multimedia. You can work with audio, video, animations, images and text. Microsoft® PowerPoint® is designed to present information to an audience and create web pages, as well as perform other tasks. The design and feel of a presentation should always match the audience's needs. Projecting a presentation is commonplace today and usually means that the work is designed for big screens. However this is not always the case. Presentation software is very powerful and is used in industry for many reasons, for example, to train and educate, to persuade people to sign up or to buy a product, or to display product information such as sales.

Multimedia presentations for users with disabilities

Audio commentary provides access to multimedia for people who are blind or visually impaired by adding sounds that describe the images, including action, scene changes, graphics and on-screen text. Captions added to multimedia presentations ensure that the audio components of the presentation are accessible to individuals who are deaf or hard-of-hearing. Captions can provide a powerful search capability, allowing users to search the caption text to locate a specific video or an exact point in a video. They are also useful for people learning to read or learning English as a second language. Commentary can assist pupils with learning disabilities by reinforcing through audio what the user is watching on the screen.

Captions and commentary may be integrated into multimedia material as a user-selectable option (closed) or permanently recorded along with the main audio or video (open). Closed captions and descriptions may be toggled on and off by the user via a preferences setting, a menu option or, in some cases, a button on the player interface. Open captions and descriptions may not be turned off (everyone sees or hears them whether they want to or not).

Working with the software

In Microsoft® PowerPoint® you will find a range of tools that help you design unique presentations. You start with a white page or blank slide. You can use ready-made backgrounds or formatted pages that allow you to work at speed on a predesigned theme. These are called templates.

Pages in presentation software are usually called slides and this is very appropriate as the projectors used are similar in appearance to the ones used before presentation software was introduced. In the past, presentations with slides were made with photographic slides and slide projectors or overhead projectors. Photographic slides had to be developed in a dark room. The slides were inserted manually into the projector cartridge and had no flexibility as they were fixed images and changing the order of presentation was laborious. Text was difficult to place on slides as it was a photographic process. Photographic slides are rarely used but do have the advantage of having very high resolution which is useful for some types of image.

Overhead projector (OHP) presentations use transparent paper-sized sheets with the presentation printed on them, slide by slide. Although animations, audio and video cannot be used, they have the advantages of being simple, they can be written on to highlight information, and are portable (they don't need a computer or special computer). Most presentation software can be used to print OHP slides.

Presentation software can be used to produce animations and web pages. There are also drawing tools.

The tools in Microsoft® PowerPoint® are very similar to those in Miscrosoft® Word with the addition of the specialist presentation tools. These include audio, animation and slide transition tools. If you are familiar with Microsoft® Word you should find the tools and dialog box layout straightforward.

With presentation software you can easily create slides that support your ideas in a visually appealing manner. Presentations can be in the form of a slide show that runs on your computer (for viewing on-screen or projected) overhead transparencies or presented on the Internet or even photographic slides.

Microsoft® PowerPoint® offers you the chance to show off your graphics skills, produce speaker's notes and handouts and include commentary music, video and animations, plus many more options to customise a presentation.

Running a presentation

When presenting information using presentation software, think about the following points that will enhance the effectiveness of your presentations.

Planning your presentation

Before you actually make your presentation, you need to think about the following questions:

- What is the purpose of your presentation?
- How can the presentation be made more interesting?
- Is there a balance of text and images?
- Will handouts be distributed?
- Is the information in logical blocks?

You may be able to save time by making use of resources that are already commercially available. However, if you are going to use such resources you must be certain to adhere to any copyright restrictions that apply.

Organising your presentation

In order for your slideshow to be most effective, it must be well constructed. It will be obvious if you have hurried or carelessly constructed a presentation, so it has to look good. Mind you, it is just as easy to overdo things, so watch out that you do not add effects just for the sake of it.

You should begin by planning your slideshow on paper, so that you know what is going to appear at each stage.

Your text should appear as a series of bullet points of around eight words per bullet point, with a maximum of six to eight bullet points per slide. Any more than this and your audience will find it hard to follow your slides.

Judicious use of snappy, concise headings will help your audience to understand the context in which the bullet points that follow are made. Likewise images can help to point your audience in the right direction and enhance the quality of your delivery.

Slideshows can be most effective if they make sparing use of:

- Sound
- Images
- Transitions
- Video
- Manipulation (using a tablet or Interactive Whiteboard)

It is easy to fall into the trap of letting the technology drive your slideshow – it can easily become gimmicky if features like the ones listed above are overused. Your audience can become distracted, confused or even bored with a badly constructed slideshow.

Top tips for slideshow design

1 Ensure your writing is not all UPPER CASE – there are other ways of emphasising important points, such as **bold**, *italic* and underlining. One of the main problems with writing that is entirely in upper case is that the reader cannot easily spot the start of a new sentence.

2 Try to use underlined text sparingly. This is because hyperlinks use underlining and although these are often coloured blue this is not always the case.

3 A sensible, visible font size is desirable. It is suggested that no writing on a slideshow should be smaller than size 24. However, it is also important that text is not too large either.

4 No audience member likes to struggle to find information on a slideshow. Therefore you must make sure that you present the information as concisely as possible. Each bullet point should summarise whatever topic you are discussing during your narration.

5 Do not overdo the use of colour. Unless you are changing background colours to suit the content of your slide, you should try to stick to a consistent colour scheme. Having said that, different colours can be used creatively to separate your slideshow into different sections.

6 Make sure your text can be seen against the background. The colour of both should contrast sufficiently to be legible. This is one point at which printing the slides is useful so you can determine what your handouts will look like.

7 Practise, practise, practise. Make sure you rehearse your slideshow before presenting it formally.

8 Produce supporting notes, do not rely on the slideshow to guide you through the content.

9 Leave time for questions from your audience.

Practising your presentation

You should familiarise yourself with the equipment you will be using. Make sure you can use the projector. If you are using a laptop ensure that you know how to display the screen via an LCD projector. If you are using an Interactive Whiteboard, make sure you know how it works – you would not want to try and learn this for the first time when you are presenting to a room full of people. It is easy to overlook the way that audio is conveyed to the audience. If you rely on an audio file in your presentation, you must ensure that you have access to speakers, otherwise your audience will not hear the audio track.

It may be that your audience has to take part in your presentation. If using an Interactive Whiteboard, it is possible to use voting units that canvas opinion from the audience which is then graphically displayed on the screen. Pace yourself as you present to your audience. Do not go so fast that your audience cannot keep up with. Do not go so slowly that you end up reading every word to your audience as they will feel insulted and patronised.

If your audience has to take notes, ensure that there is enough light in the room to do so. Modern LCD projectors do not require a complete blackout in order to see the presentation.

Using video in a presentation

Videos can be embedded within slideshows, but these should be used with caution as doing so can slow down your presentation, affect the way it is printed out and increase the file size dramatically. One way to do this efficiently is insert a hyperlink to an external location where the file can be found, rather than embedding the whole file. This requires you to have a live Internet connection as you carry out your presentation, and this may not always be available.

Evaluating

After planning and organising your presentation, you need to review it and consider revisions to help you accomplish your objectives more effectively. Are the most important points emphasised? Is the order of content presented logically? Is there an introduction and conclusion? Would it benefit your audience to have a handout of the slides prior to the class?

Using a video in a presentation

You can sometimes find video on the Internet. Once you have found a movie clip ask yourself: 'How can I include this movie in my presentation?' Generally the most straightforward answer to this question is to insert a web link into the presentation. The steps for this in Microsoft® PowerPoint® are:

1. Copy and paste the URL of where the web movie is located, or simply make a note of it.
2. In Microsoft® PowerPoint®, use the AutoShapes button in the Drawing toolbar to select an Action Button. Select the Movie action button.
3. The screen cursor changes to a cross. Hold down the left button of the mouse and drag the cross to the desired size rectangle for the button. The Action Settings dialog box opens when you release the button.
4. Select the Mouse Click tab and tick the Action on Click Hyperlink to radio button. From the drop-down list select URL... . Paste or type in the URL of the movie web site (see step 1) in the Hyperlink to URL dialog box.
5. When you view the presentation, click on the Action Button to launch a new web browser window and open the webpage containing the video clip you want.

The big disadvantage of this technique is it requires direct access to the Internet during the presentation. If Internet access is interrupted or is slow playback of the clip will be directly affected. Neither does playback of the clip take place within the presentation software. This means the movie clip will not fit seamlessly into the presentation. Despite these disadvantages, using a hyperlink within a presentation to a web movie can be an effective and fairly simple way to include video within your presentation.

The following table gives some alternatives.

Alternatively, movies can easily be directly inserted into a multimedia presentation, but a *local version* of the file must be obtained first. However, to protect copyrighted content, many movie files on web pages cannot be directly saved by users.

Movie files that can be directly saved to your hard drive have *direct movie links*. The file extensions of these links are not the usual .htm, .html, or .asp extensions. Direct

Option	Internet access required?	Advantages	Disadvantages
Hyperlink to Web	Yes	• Easy and fast. • Content up to date.	• Requires reliable Internet access. • Not 'seamless' especially with dial-up access. • Content can change or even disappear.
Save and insert local copy	No	• Reliable. • Files can be stored at high resolution.	• Many Web movies cannot be downloaded and saved. • Large file sizes.
Screen capture	No	• May be the only way to include an offline copy of a Web movie.	• Time consuming. • Requires specialist software.
Digitise a movie	No	• Give the most control over properties and quality.	• Time consuming. • Requires specialist software.

movie links have the file extension corresponding to the type of compression format used in the video clip, for example .mov, .wmv, .mpg, and .rm.

Once you have found a direct movie link you can right-click the link and save the linked file to your desired location. It is usually a good idea to save the movie file to the same folder where the presentation file is saved. An important thing to note about inserting movie files is that large movie files can slow down presentation performance.

If 'live' Internet access is not available during a presentation and a direct movie link to a video file cannot be found, then screen-capture software, can make even these web movies 'save-able' and 'insert-able'. The steps for using screen-capture software to save an online movie are generally similar.

Generally the faster and more powerful your computer is, the smoother and better quality the captured video and audio will be.

The movie file created by a screen-capture program may be large, however it can be reduced in size with different programs. QuickTime Pro® is available for both

Microsoft® Windows® and Mac® users, and allows video files to be opened and exported in a wide variety of formats. Windows® media file video clips can be imported and sequenced with other video file formats and then exported as a single movie file. That file can subsequently be inserted into a presentation as described earlier in this section.

Sometimes, a video clip that you want to include in a presentation is not available online. It may be part of a full-length movie available in VHS or DVD format. As mentioned earlier, it is important that you thoroughly understand the copyright issues related to copying this material. Assuming the proposed use of the desired video content constitutes 'fair use', there are several viable options.

One method is to purchase hardware that connects to the video playback device (VCR or DVD player) and your computer. These devices allow video to be 'digitised' (although technically DVD video is already in a digital format) and made into shorter movie clips. These hardware solutions can take the form of a capture card installed within your desktop computer or an external capture device that plugs into a USB or FireWire® computer port. If you already have a digital camcorder you may not need an additional piece of hardware to capture video from VHS or DVD. By plugging your camcorder directly into the video playback device, you may be able to record a desired video segment directly. You can subsequently import the taped segment into your computer using free software that comes with your computer's operating system. Digital camcorders can often be used as direct 'line in' converters for video sources as well. If you can connect your camcorder to the video playback device (usually with a three-part cable: yellow for composite video, and red/white cables for stereo audio) along with a FireWire® cable to your computer, you may be able to directly import video from VHS and DVD to your computer's hard drive.

Using Microsoft® PowerPoint®

Microsoft® PowerPoint® is probably the most popular of several software presentation programs. Originally developed for business purposes, it has quickly penetrated education as a valuable classroom tool. It is flexible and you can develop professional-looking presentations from the most basic of text transparencies to multimedia programs with audio and video. Microsoft® PowerPoint® presentations can offer visual appeal, up-to-date links to the World Wide Web and the ability to address various learning and teaching styles. Audiences will also appreciate the organised format that Microsoft® PowerPoint® adds to lectures through outlines, notes and handouts. When used effectively, this tool can emphasise key concepts, stimulate interest and promote understanding.

Starting a new presentation

Starting a new Microsoft® PowerPoint® slideshow is straightforward because the first dialogue box you will see gives you a set of choices from which you can select the most appropriate option.

The options are:

- Autocontent Wizard – this guides you through the creation of a slideshow step by step. To work through this, simply read each box you are presented with, alter any details that need to be changed and click Next until finished.
- Template – this gives you a background template which can help to make a consistent slideshow.
- Blank presentation – this is the best option if you want the freedom to create a slideshow that is entirely to your own specifications. But you probably shouldn't try to use this until you gain confidence.
- Open an existing presentation – this lets you continue to edit a slideshow that has been saved previously. It is possible to bypass this step by double-clicking a Microsoft® PowerPoint® presentation directly from where your files are stored.

After choosing your preferred option click OK.

Saving a presentation

To save a presentation, simply choose Save or Save as from the File menu or press Ctrl+S.

If you are unclear about what to do at any stage of the slide creation process you can make use of the Office assistant to provide help. You can type questions in natural text, or else use keywords.

Pressing F1 is a speedy way to generate a list of Help topics.

Using Microsoft Powerpoint

Viewing your presentation efficiently

There are several ways to display a slideshow. These can be selected by clicking on the tiny buttons at the bottom left hand corner of the Microsoft® PowerPoint® window.

1. Slide view – shows one slide at a time. This is where you would develop the formatting of each slide.

2 Outline view – shows all text that is included as bullet points. This is the section in which you can create the textual content of your slideshow.

3 Slide sorter view – this shows all of your slides as a series of thumbnail images. This can be used to change the order of slides or to delete and copy and paste slides.

4 Notes page view – this is used to create a set or presenter notes to accompany each slide.

5 Slideshow view – this presents your slide as your audience will see it.

6 Normal view – this feature is found in Microsoft® PowerPoint® 2000 and subsequent versions. It replaces the Slide, Outline and Notes page views by combine all of them onto the one screen. The borders of each pane can then be adjusted to suit your own needs.

Zooming in and out

Sensible use of zooming allows you to see the required level of detail in your slideshow. The magnification is determined by selecting from the drop-down list on the toolbar. As well as preset magnifications, you can type in your own zoom factor.

Finding a presentation file

The following procedure can be used to find a file if you have forgotten its location. The same procedure can be used to locate any files not just Microsoft® PowerPoint® ones.

On a PC

1. Click Open in the main toolbar
2. Click Preview in the Open dialogue box toolbar – this shows the contents of the file.
3. Use the Look in drop-down list to choose the drive where the file should be located.
4. Click Commands and Settings and then choose Search Subfolders.
5. Complete the criteria for the search that you wish to carry out e.g. if you know the file contains the word 'egg', type egg in the Property box.
6. Click Find Now.
7. Select your file from the list presented. If the file does not appear, then it either does not exist, does not contain the text you typed in at step 5, or exists elsewhere.

On an Apple Mac

1. Click Open from the main toolbar
2. Click Find File
3. Chose the location of the drive you want to access
4. Click Advanced Search
5. Complete the criteria for the search that you wish to carry out e.g. if you know the file contains the word 'egg', type egg in the Containing Text box.
6. Click OK twice.
7. Select the file you want to open from the list that is presented.
8. Click Open.

Using Microsoft® PowerPoint®

Building a presentation

Adding slides

A slideshow is built by adding slides. You therefore need to ensure that the template of the slide you load is suitable for your needs. To add the most flexibility, you might want to use a blank slide although this can be more time consuming.

In the New Slide dialogue box you can select Auto layout with the layout that you want, but once you gain confidence you may find that the options you are given are a little restrictive.

Choose the layout you want and click OK.

Applying a background

There are several options for backgrounds:

- You can choose a single colour
- You can choose a graded selection of two colours
- You can choose a pattern made from a combination of two chosen colours
- You can add your own picture as a background
- You can choose from a preset series of background designs such as denim, paper bag, marble etc.
- You can choose a design template – but you can only have one of these per slideshow.

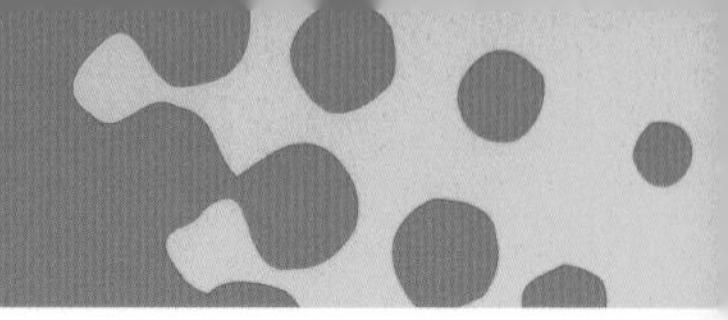

To choose the latter, select Apply Design from the main toolbar or choose Format > Apply Design .

Developing your outline

This is a great way to develop your slideshow in a methodical way.

Importing an outline

You can design an outline in Microsoft® Word® and then import it to be the basis of your Microsoft® PowerPoint® slideshow.

You do this by giving:

- main headings Heading 1 style,
- bulleted lists Heading 2 style
- second level bullet points Heading 3 style

From here you can create a slideshow by choosing File > Sent To > Microsoft® PowerPoint®. This is a well-kept secret that illustrates the way in which several software packages can be used in conjunction with one another.

How to get good marks

✓ **You need to show that you have a good understanding of the potential of this kind of software for creating multimedia products.**

Homework

1. Using a microphone. load up a sound recorder and save a sample of your voice, load up Microsoft® PowerPoint® and insert the sound file you have saved. Play the slide and click on the sound file icon.
2. Design a slide using a serif type face then copy it and change the font to a san serif face. See if it makes any difference
3. On one slide add as many multimedia objects as you can! Sound movies, animation of text and objects, images. The result will probably be difficult to follow but this will help you learn how to put in different elements and think a bit about design.

Website software

What you will learn in this chapter

You will learn how to create a simple webpage, using text, images, animation and video. You will also find out about file sizes and optimisation, buttons and rollovers.

This chapter is concerned with developing an online presence using appropriate software to produce websites.

When designing and making a multimedia product, you will need to balance many conflicting aspects. You will also need to consider a whole range of rules.

Before you can even start to design a web page you need to have content. What will your page have on it? You don't need to be precise at this point, but you need some idea.

Earlier chapters have mentioned the need for gathering resources: web page design is an area where you need to start early. You may need to collect text from one source and images from somewhere else. You may also need to generate your own movies and audio tracks.

Chapter xx explained the structure you could consider for storing your material. When using web page design software, you will need to be very organised because the links that are generated between pages will only work if the web structure stays unchanged.

Load time

An important thing to consider when developing web-based multimedia products is the length of time it takes for a page to load. If what you produce takes so long to load that the user clicks away from it, it doesn't matter what else you have done because nobody will ever see it!

This aspect of site design is called 'optimising'. It is the operation of making sure that everything on your page works well together and loads quickly.

Most Internet users will only wait a few seconds for a page to load on their screen. The average wait time is 8 seconds. This means that if your web page takes more than 8 seconds to load, the average user will start to click their mouse and move to another page. One way around this is to have things loading quickly to keep their attention, even though the whole page takes a little longer. The load time will of course depend on the specification of the user's computer and the speed of their Internet link. Try your site on different systems.

There are a number of websites that offer to check your load times. Most professional sites use some sort of optimisation software, and are therefore able to keep their load times very low.

You should consider the potential load time at every stage of the design.

Designing a page

Web pages are very simple to make. They can be generated using a word-processor application or a graphics program and then saved in an appropriate format for viewing through a browser.

There are a number of professional web design applications if you want to specialise in this area: Macromedia® Dreamweaver® or Adobe® GoLive®, for example.

Many web designers consider layout to be the most important aspect of design after the load time. When you use a web design application the first thing you are asked is what layout you will be working with.

Whatever you design it must be clear and easy to understand. If you look around on the Internet you will soon find pages that you can understand and others that are confusing and poorly laid out.

The way most Western people see the world is the way they read – top to bottom, left to right. This also applies to websites, so the most important information should be at the top, the left should be for secondary information and the bottom right for the general material.

If you look at newspapers, magazines and web pages there is a common feel to the way they look.

Professional designers of newspapers, magazines and web pages have been trained how to properly lay out material. You can learn from things around you.

- If you want to design a page that carries lots of text, like a newspaper, then you should look at newspapers and similar publications to get a feel for the ways that they communicate their information.
- If your page is mostly images, then look at catalogues and magazines to see how they work.
- If you are developing pages that have video or audio, watch the television. Many programmes have particular 'page layouts', (e.g. news channels may have running headlines across the bottom of the page, credit pages generally run right to left or bottom to top, etc.)

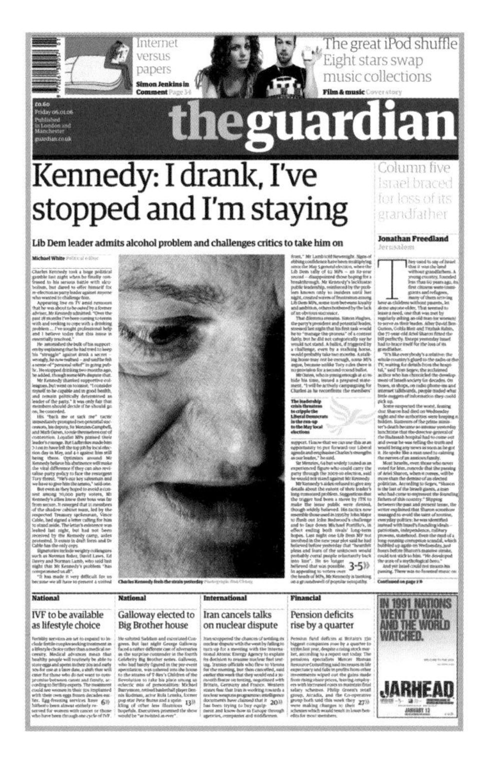

Internet versus papers
Simon Jenkins in Comment Page 34

The great iPod shuffle
Eight stars swap music collections
Film & music Cover story

£0.60
Friday 06.01.06
Published in London and Manchester
guardian.co.uk

theguardian

Kennedy: I drank, I've stopped and I'm staying

Lib Dem leader admits alcohol problem and challenges critics to take him on

Michael White

Charles Kennedy feels the strain yesterday

Column five
Israel braced for loss of its grandfather

Jonathan Freedland
Jerusalem

Continued on page 2

National
IVF to be available as lifestyle choice

National
Galloway elected to Big Brother house

International
Iran cancels talks on nuclear dispute

Financial
Pension deficits rise by a quarter

IN 1991 NATIONS WENT TO WAR AND THE WORLD WATCHED.
JARHEAD
JANUARY 13

Pages that carry a mixture of text and images are the most difficult to design well. Again, for inspiration look through magazines and other publications.

Designing your layout should be done on paper. Most designers are pretty good at art, but it is not essential.

When you start to transfer the idea to the computer, there are lots of helpful guides and templates that you can use.

When you have some idea of layout, you need to get working on the content.

Content

The way the site files are stored and operate can have an influence over the speed that your pages load.

Within your main folder you need at least two subdirectories:

- Images, where you will keep all the pictures that will be used in your site.
- Pages, where all individual pages will be stored.

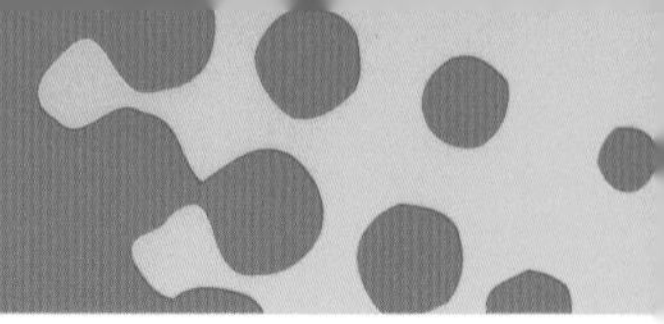

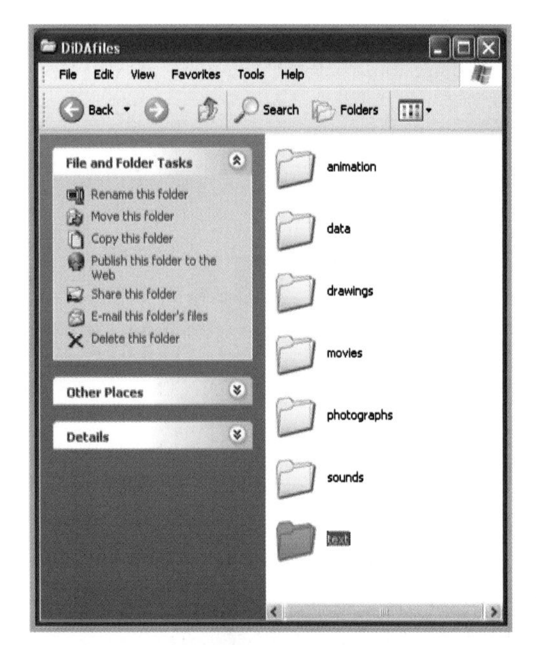

But, as you are probably going to want to involve a number of other elements, you should consider using more folders.

In the first book in this series, *using ICT*, a simple, 'static' website was developed using Macromedia® Dreamweaver®. You will have used a storyboard to plan each page.

Here are some ideas on how to add some multimedia elements. Adding video and audio to a basic HTML page can be difficult because HTML is designed to show text and images. The following instructions show how to add these video and sound to a simple Macromedia® Flash® page and then add that to a HTML page.

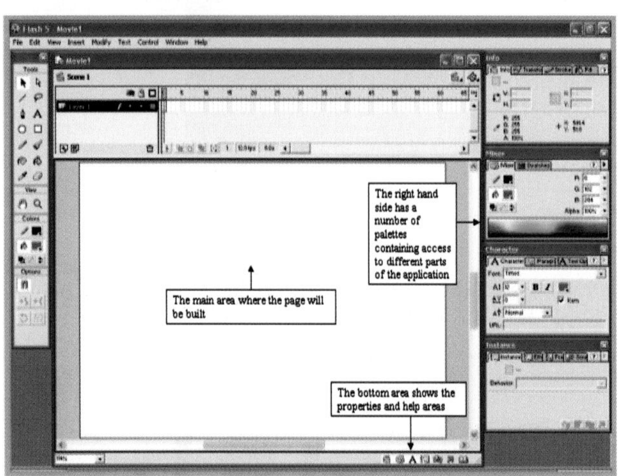

Flash® can be set up in a number of different ways. Here we have used the default design view.

When working with Flash®, the software is set up to build a multimedia product. You have already set up a simple folder structure and you will now use it as you build your product. The same structure will allow you to build a fully functioning multimedia website.

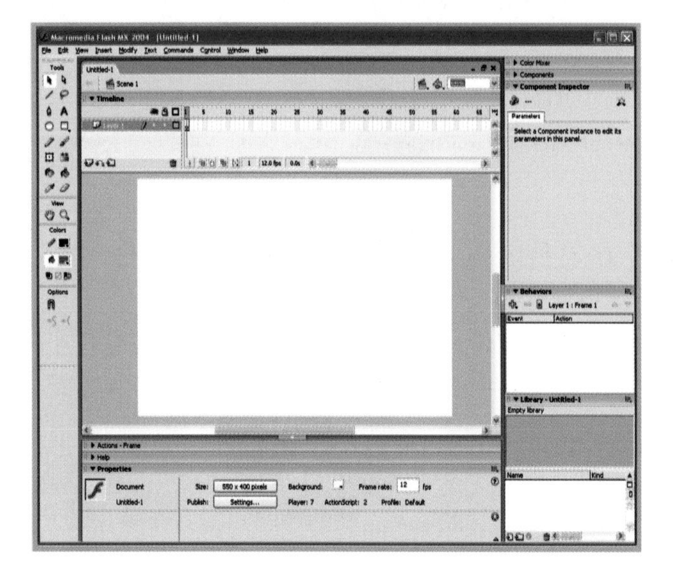

Flash starts with a blank page. This will appear as a white background, the top of which holds the page title (as it has not been saved yet, it is 'untitled-1').

1. Save the blank page in your root folder, not your animation folder. Save your page as *animation01.fla*. The new title should appear on the tab at the top right.
2. The properties inspector below the page changes as you work, so make sure that you start off with the correct properties section. Move your mouse pointer to the middle of the page and right click. From the shortcut menu choose document properties. This dialog box allows you to set various properties for the

page. Later we will set properties for elements that will be appearing on the page.

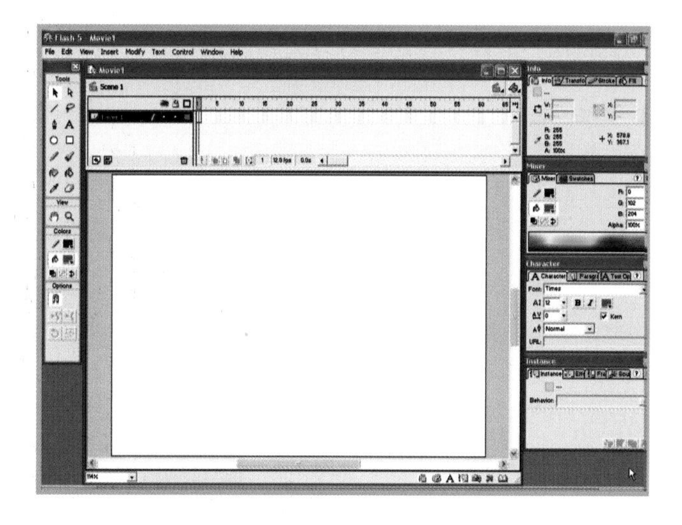

3. Set the appearance of the page: Set the size of the page. It should be smaller than the web page you will be adding it to. In this example it is 550 × 400 pixels, easily small enough to fit on a standard web page. Choose the background colour. Think about what you will be adding to the page and select a colour that is suitable. The other settings can be dealt with later, so click OK.

4. To get the text in the right place, the page needs to be arranged correctly. Typing straight onto the background is the easiest way to build your page, but it can create difficulties later, so place the video first. It can always be removed later. On the File menu, click Import, then navigate to the folder in which you have stored your video. Select the video you want to place on the page and choose Import to library.

   As video can be stored in a variety of formats, Flash® will try to recode the file into a suitable format.

   Your application may offer different options, but the principle will be the same. You will be able to add the video to the library or edit it and then add it to the library.

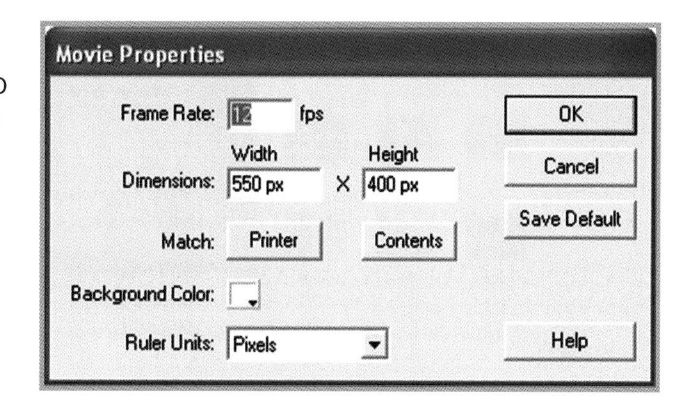

5. As you import the video, you will be asked to select compression rates. Choose the lowest possible. If it is not good enough, you can always import it again! When the compression has taken place you will have the file in your Flash® Library.

6. To add the video to your Flash® page, simply drag it from the library and place it on the page. You will be asked if you would like the timeline to be extended to match the length of the video. You will probably select yes!

   The video will also have a layer, next to the timeline. Rename this *video1*.

7. Now to add some exploding text! Use the text tool to enter your title. When you are happy with the text content, click on the selection tool and move it to where you want it to start from. Use the property inspector to adjust font details.

8 As each letter is going to move separately, the word needs to be broken apart. Go to Modify > Break Apart.

9 To ensure that the letters move as planned, each needs to be converted to a symbol. Right click on each individually and choose convert to symbol.

10 Choose movie clip and make sure the registration is central.

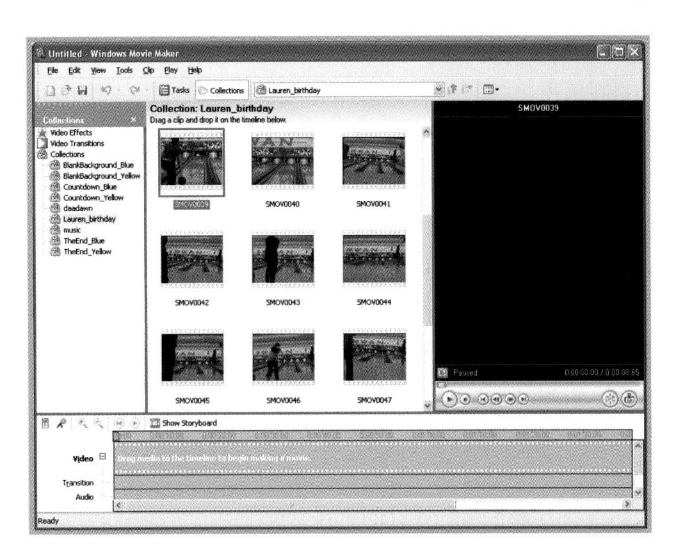

11 Give each letter a different name. If there are duplicate letters you only need to name one of them to save file size.

12 Highlight all of the letters and right click distribute to layers. Each letter (symbol) will now have its own layer with a name, as well as a place in the library.

13 On the timeline, go to the last frame that has content and enter a keyframe into the last cell for each of the letters (F6).

14 Go back to the first frame and press return on the keyboard. The movie should run, but the text should stay as it is.

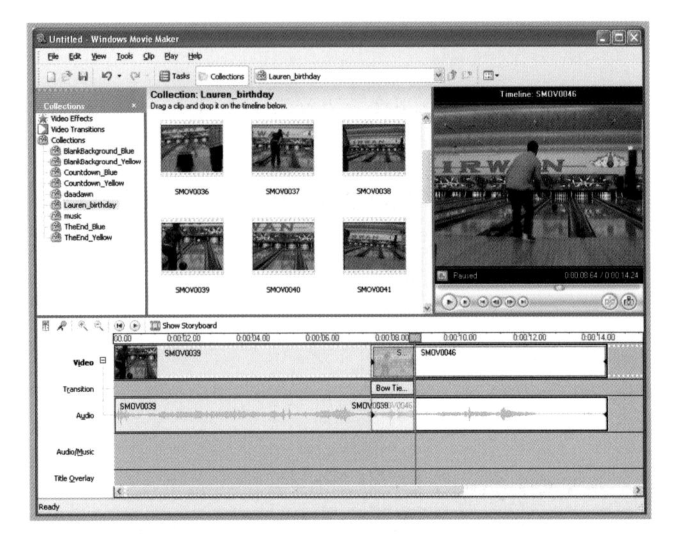

15 Select a frame number and add a keyframe to each of the letter layers.

16 Move each letter to a different location, on or off the page.

17 Click in the first frame of a layer, hold down shift and click in the keyframe of the same layer. This will highlight each cell. Right click and Add Motion Tween. Do this to each layer containing a letter.

18 Go back to the first frame and press return.

19 To make the letters bounce around more, add more keyframes and move the letter to a new point on the screen.

You can also experiment with rotating the letters at each keyframe or editing the properties.

To add a soundtrack you need to import a suitable music track in the same way as the video.

1 go to File > Import, and navigate to the sounds folder. This allows you the option to add it to the library.

2 Create a new layer. Click on the 'new layer' icon on the timeline menu.

3 Drag the sound from the library onto the stage. The sound profile will then be added to the timeline, try to line up the peaks of the sound with the keyframes of the animation for a really good animation.

4 When you are happy, save the file. This saves all of the information ready for editing at a future date.

5 To add it to an HTML page, you should export the file. Go to File > Export > Export Movie.

6 Give it a name (different from the .fla name). The Flash Player® dialog box will then open allowing you to choose the format of your movie. Unless you have a particular reason not to, go with the defaults. Now to add your animation to a webpage.

7 Open a blank HTML page in Macromedia® Dreamweaver®, or use an existing page.

8 From the Common toolbar, choose media plug-in and navigate to your animation. If the animation is stored outside the root folder of the website, you will be asked to copy it to a suitable location.

9 Adjust the size and position of the animation. It may be best to locate it in a table. When you are happy, preview it (F12)
and save it with a name you can remember (if it is to be a welcome screen, call it *index.htm*).

10 Add any other detail, but remember that as with any material use only what is required to get your message across to the desired audience. Animation, video, sound and other embellishments can be off-putting in certain professional contexts!

How to get good marks

✓ **You need to show that you can create a web page that looks good and has a range of features, including graphics, sounds, video and animations.**

Homework

1 Produce a welcome page for a website. There are a number of DiDA sites available for you to get inspiration, try http://dida.edexcel.org.uk/home/ or http://dida.nwlg.org/ for example.

2 Create a multimedia web page that has synchronised movement and sounds, such as gun shots making holes in a wall.

3 Create a library that can be shared across a number of Flash® movies, including text, video and sounds.

Word-processing software

What you will learn in this chapter

In this chapter you will learn the basics for using Microsoft® Word. You will read that Microsoft® Word can do a lot more than write letters and booklets. You will learn how to format text, and draw and create shapes for diagrams. You will see that Microsoft® Word can be used to produce a simple web page.

Most word-processing software can be used to create multimedia products. Microsoft® Word, for example, is capable of showing movies, animations, images and of playing sound. Within Microsoft® Word are drawing tools and a range of clip art and photos. It also has tools for creating web pages and image editing facilities.

When we talk of word-processing software we really mean software that we can write with. Microsoft® Notepad is a basic text editor that allows only simple formatting. Microsoft® WordPad is a simple word processor, but Microsoft® Word allows you to do much more. Word processors replaced typewriters that relied on accurate typists and did not really allow for mistakes. You could not put in images or change the range of fonts.

Microsoft® Word has evolved from being a simple word processor and now you can even use it to make web pages. You can use it for many types of multimedia presentations, however you still need to use desktop publishing packages for professionally laid out material.

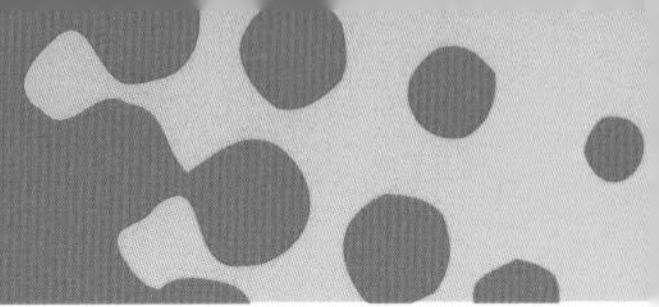

Using Microsoft® Word

Working with graphics

It is possible to do so really good graphic work in Microsoft® Word using the drawing, painting, image manipulation and object manipulation tools.

You can insert several kinds of graphics (e.g. a graphic file, clip art or AutoShapes) into a Microsoft® Word document.

Inserting a graphic from another location

1 Choose Insert Picture > From File... .

2 Find the file and select it.

3 Click Insert.

Tip

To more easily move a graphic file, insert a text box first. With the text box selected, insert the file.

Inserting clip art

1 Choose Insert > Picture > Clip Art... . You may be prompted to insert your Microsoft® Office CD-ROM for more clip art.

2 Choose a category, and then choose a picture.

3 Click OK.

Inserting AutoShapes

1 Click the Drawing button on the Standard toolbar this will display the Drawing toolbar, if necessary.

2 Check AutoShapes.

3 Choose a category and then click the shape you want.

4 Click inside your document where you wish to add the shape. You can also drag to set the size as you insert or modify it afterwards.

Tip

You can click Import clip to add graphic files to the clip art gallery for easy access.

Modify graphics

Click the image to select it. You can then resize it using the image 'handles'. You can also use the Picture toolbar that appears to modify the image in several ways, including cropping, brightening, etc.

Modifying an AutoShape

1 select it and use the 'handles' to resize it.

2 Drag it to move it. (You'll see a four-arrow cursor.)

3 Press Del to delete it.

4 Choose Free Rotate on the Drawing toolbar to rotate the shape. Drag the green circles until you have the rotation you want.

Choose Fill Colour to fill the AutoShape. You can choose a colour, choose More Colours for more options, or choose Fill Effects to fill the shape with a gradient, texture, or even a graphic image.

Choose Line Colour to set the colour of the line around the shape. If you fill the shape with a colour, you can remove the line (choose No Line) if you wish. Choose 3-D to create a neat 3-D effect.

Choose Shadow to add a shadow effect.

Use the Draw menu on the Draw toolbar to rotate or flip the shape and for other options.

Formatting

Creating a numbered list

Making a numbered list is easy. Here's how.

Type each item without any number, full stop, or space – just the text itself.

When you're done, select all the items.

The second line of the text should not go back to the left margin. A bulleted list is formatted just like a numbered list, except that it uses bullet points instead of numbers.

How to create a bulleted list:

Type each item without any bullet symbol or space – just type the text itself.

When you've done this, select all the items.

Click the arrow to the right of the Font Colour button at the right end of the Formatting toolbar (it has a letter 'A' on it).

Deselect the text to see the result.

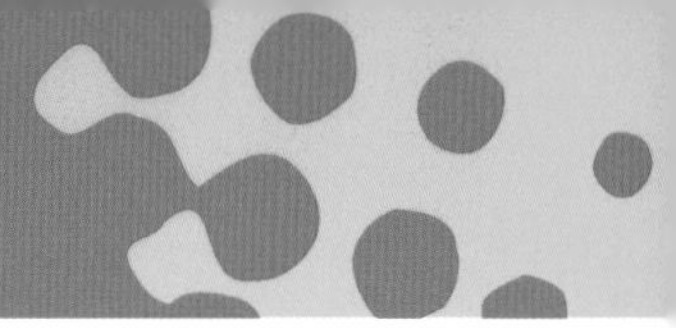

Copying formatting from one place to another

An easy way to format text is to copy the formatting from somewhere else that has the formatting that you want to use. Formatting include things like font, font size, font colour, bold, italic, underlined, and so on. Here's how you do it:

1 Select the text that has the formatting you want to copy.

2 Select the text you want formated in this way.

Tip

To format several different sections of text with the same format double-click the Format Painter button. Select the text that has the formatting you want, and then select all the sections of text that you want to format. Double-click the Format Painter button again (to turn it off).

Aligning text

Highlight the text and click an alignment button from the Formatting toolbar: align left, center, align right, justify. (You can click an alignment button just before you type text to make sure it is aligned in advance).

Indenting text

To indent a whole paragraph, select it and click the Increase Indent button on the Formatting toolbar (→). To reduce indentation, select a paragraph and click the Decrease Indent (←).

Inserting a page break

1 Click where you want the new page to be started.

In the Break dialog box, click OK.

If your document has columns, you can choose Column Break to force the text to start a new column.

Vertically centering text on a page

You can vertically centre all the text on a page. This is useful for creating title pages.

1 Choose File > Page Setup… .

From the Vertical Alignment drop-down box, choose Centre.

In the Document dialog box, click the Layout tab.

Click OK.

Changing line spacing

1 Highlight the paragraphs you wish to change the line spacing of.

2 On the Indents and Spacing tab, choose the line spacing you want from the Line Spacing drop-down list.

Adding page numbers

To add and view page numbers, you must be in Page Layout view. Click the Page Layout button at the bottom left corner of your screen.

Choose Insert > Page Numbers.

If you don't want to display a number on the first page, untick the Show Number on First Page tick box.

Click OK.

Adding headers and footers

To add and view headers and footers, you need to be in Page Layout view. Click on the Page Layout button (this will be at the bottom left corner of your screen).

Choose View > Header and Footer. The Header and Footer toolbar appears.

To create footer, click the Switch Between Header and Footer button on the Header and Footer toolbar.

Choose Insert > Footnote.

Choose either AutoNumber (to number footnotes consecutively) or Custom Mark (click Symbol to choose a symbol).

Click OK to place the footnote.

Unfortunately, you have to return back to where you started yourself.

Adding footnotes is very straightforward as the renumbering process is automated if footnotes are added at earlier points in the document.

Autoformat as you type

Some people dislike this automated process as pages can often look unusual and can be presented in ways that is not what the user intended at all. In fact it can end up taking a great deal of effort to undo the automated formatting that the software has applied. This happens especially with bulleting and numbering and borders if you type

three or more underscores, hyphens or equals signs and the enter key a border is generated.

Using styles to format text

Styles can be used to good effect within documents which are more than a couple of pages in length. Although a range of Styles are embedded within Microsoft Word®, you can create some of your own. One of the main benefits of using styles is that if you want to apply a new style to the way your headings are displayed, you can do it in one step, rather than having to be repeated for each occurrence of the heading.

If you want to change the line spacing or other factors affecting the way text is displayed, you can do so using Format > Style.

If you create your own new style, then you should give it a distinctive name so that you can identify it later. You can do this from the dialogue box by clicking OK. Then click Apply to enact the changes on the text. You can also alter the number of text columns within this box. By choosing 'This point forward' you can ensure that the number of columns will apply from nopw until the end of the document.

To change page margins on a PC

Choose File > Page Setup

On an Apple computer:
Choose Format > Document

Creating and using tables

Tables can be used to display text and data in a structured way. This can be achieved by choosing Insert Table.

Rows can be added by clicking below where the row is to be placed and choosing Table > Insert Rows.

Adding a row at the end of the table is slightly different because you will need to click the bottom right cell and press enter.

Mail merging

You can make personalised letters and other documents by merging data held on a database with documents that have predefined fields embedded within them.

Creating the database

In a new document go to Insert > Table. This needs to have at least two rows and as many columns as you need to fit all your fields in.

In the first row you type your fieldnames and in the second row the data for the first record is entered under each relevant heading.

Then choose Tools > Mail Merge to open the Mail Merge Helper. You may not normally wish to make use of help facilities within software packages but this is one instance where it is advisable for you to do so the first time you use this feature.

The Mail Merge Helper will ask you if you want to use the active window or create a new document. If you have followed the steps above the document will be on your screen so select Active Window.

Getting the data

After you specify the active window click Get Data.
Choose your data source and click Open.

Inserting database fields into your main document
Follow these steps:
Click Edit Main Document and you will be returned to your main document. If you need to input a space, or punctuation after any of the fields (such as the name after 'Dear' at the start of a letter) then it should be added at this stage.

For labels, click OK to close the Create labels dialogue box.

Merging the main document with the database
Click Mail Merge Helper on the Mail Merge toolbar. Then change the settings as required. Microsoft® Word® will create a new document with all the records merged into the main document. Automatic page breaks ensure that new documents start on a new page.

Then in the Save As Type list, choose Document Template and then click Save.

The document can be presented in four ways:

- Normal view – this shows plain text without margins, headers, footers or page numbers. This can be very misleading and despite its name it does not show what the page will normally look like!
- Online layout view – Microsoft® Word® displays the document map on the left in order for you to jump immediately to specific sections of the document if necessary.

- Page layout view – shows the document as it will appear on the printed page. This is the most commonly used view within Microsoft® Word®.
- Outline view – is used to show the structure of the document by featuring the main headings only. This is useful if you do not need to see the majority of the text in the document.

Zooming in and out

As with other software programs, you can use the zoom facility to look at aspects of a document in detail, or to see the whole page on the screen.

How to get good marks

✓ **You need to be able to demonstrate that you have a good understanding of word-processing software and the range of features that can be used to enhance your work.**

Homework

1. Design a Microsoft® Word page with hyperlinks to a website, one animated clip art and one sound file. Save it as a web page and view it with a browser.
2. Using the drawing tools and AutoShapes, draw a plan of your house and garden. Fill with the fill tool.
3. On the Internet, find an image and drag it into Microsoft® Word® using the mouse. You will need to use the restore down button on the top right hand side of your screen for both web and word.

Summative project brief

The final marks you are awarded are based on the Summative Project Brief (SPB). The SPB will probably come to you on a CD and your teacher will explain what you need to do.

This is the project that the exam board send to you in the final year of the course. This also is the part of the course on which you are awarded your grade. It is a focus for all the skills and knowledge you have acquired during your DiDA course!

The SPB asks you to solve a problem by using these skills and knowledge to produce a piece of work.

It is recommended that a minimum of 30 hours should be spent working on the SPB. Your teacher may allow you a little more time. It must be clear to the board that you have done the work yourself and that the work was supervised by a member of staff qualified to do so. The final product must be your own work.

Your SPB could be built up in the following sequence, perhaps as a series of mini projects:

- Getting information together
- Planning
- Researching
- Making a database
- Market research
- Graphics
- A presentation and a final report.

During this stage you need to save your work carefully and keep a backup. If you lose your work it cannot be marked. If it is saved and is difficult to retrieve you will lose marks. If you fail to complete a part you will lose marks!

There are details about the e-portfolio elsewhere in this book and there is information available from Edexcel. Make sure you check the details with your teacher.

Create a folder for your project and then create sub-folders for specific types of work, for example graphics, databases, research, etc.

Practice project

You are required to produce three items that can be used to encourage prospective parents to send their children to your school. The items should be:

- A trailer to be sent to local primary schools. This will be shown to the pupils before their parents visit the potential secondary schools.
- A virtual tour to be viewable on the Internet via standard browser technology.
- A game which should be an on-screen activity that Year 6 pupils could take part in after they and their parents have visited the school.

All three items must total less than 15 MB.

When producing these products you should remind yourself of the audience and the purpose. Not everyone taking part in the project has English as a first language, so it will be inappropriate to rely too much on text or speech.

To ensure that your audience enjoy looking at your material, you need to ensure that they work well. This means using prototypes throughout the development of your products and testing them with users.

If your products are effective, you will have demonstrated your ability to apply your ICT skills in the best possible way.

Before anything else, prepare a Work Plan which addresses the questions:

- What information do I need to gather?
- How can it be presented?
- Who will it be presented to?
- Where will the information be presented?
- What software would help me?
- How can I store the information?
- What will I need to do first?

Your plan could be produced as a single page with hyperlinks to various files or folders.

What information do I need?

Think carefully about the sort of information you would have wanted to know when you moved to another school.

You also need to collect information about your school such as rules, uniform, activities, trips and other interesting data. You may need to interview some of the staff or pupils.

You need to gather a variety of multimedia components for use in your product. You must produce some of the components yourself and use some that are ready-made. These components could include video clips, audio clips, text and images. You may be able to use a component in more than one product.

You will need to decide how best to present the information to make sure that the audience is able to understand it and be interested enough to want to learn more.

How to present the data

You need to enter the information into an application that will allow you to manipulate it to produce the three items. Bear in mind that it should be a set of multimedia products, so you should consider how best to show each component.

Where to present the data

You will have access to a range of facilities to test your products, but most of it will need to be displayed on a screen, through a standard browser.

What software is available?

You need to check with your teacher if you are unsure, but you should have access to Microsoft® Office applications (or similar) and image manipulation, sound recording, video editing, web design, animation and other software.

What's first?

Start by planning what you are going to do and developing a timeline. Build in time to collect components and generate your own resources, as well as manipulating them and combining them for the products.

How can I store the information?

Your teacher will allocate an area on the school system for you to save your work, but you will have to set up a file or folder structure. Remember to keep regular backups.

Hints and tips

Remember a picture paints a thousand words. Use images and video where you can to illustrate a particular idea. Don't rely on the audience reading text.

If something doesn't **add** to your product, it probably **detracts** – and nothing distracts like unnecessary effects.

Do not use fonts that are complex, use simple, easy to read fonts. Some experts recommend a font designed specifically for online viewing, such as Arial® but others would choose readable classic fonts. Remember that ornate or very blocky fonts are difficult to read on-screen.

Effects such as Microsoft® WordArt have their place, but less is generally more in a professional presentation. Typography is a skill. Professionals carefully choose or create fonts.

Choose colours carefully. A Black or dark blue background with white or light yellow text is easy to read. Some people prefer black text on a white background (although the white background is somewhat harsher, it is easier to see in a room with more light).

If you add background designs, keep them simple, and make sure they don't overlap and obscure your information. Consider the quality of the screen you will be using. Beautiful backgrounds and colours may not look the same on a low resolution screen.

Do not beat about the bush! Remember that your audience is busy and intelligent.

Check for unintended colour combinations or symbols. Likewise, look for incongruous background shapes, like shapes that look like stop signs on an otherwise positive presentation.

Evaluation

When you have finished your multimedia products, review your work. How well does it meet the needs of the audience? Is it fit for purpose? How could it be improved?

You should evaluate your work throughout the project. How well did your plan work? What did you learn from the prototype? What went wrong and why? Is there anything you would do differently?

You should keep a record of the feedback you received whilst working on the project. You need to use that feedback to inform your choices. Record this information in your Quality Log.

Your evaluation could be a written report, but you could also make use of video clips, audio clips and presentations.

Do you want top marks for your projects?

If you want to attain the highest marks:

- Look at the SPB regularly and stick to it.
- Remember your aims and stick to them.
- Test, check and evaluate regularly.
- Ask your friends and family to look at what you have done and ask for feedback.
- Check your timeline or plan regularly.

If you want to lose marks!

- Never refer to your timeline or planning notes.
- Research poorly or inappropriately.
- Ignore feedback or constructive comments.
- Fail to include elements from each software group.
- Make your work difficult for others to read.
- Have a weak, or not evident, evaluation process.

Analogue – signals which vary continuously, for example, temperature readings from a sensor.
Application – A piece of software which fulfils a particular, and often specific, task; examples are spreadsheets, word processors, graphics software.
ASCII – American Standard Code for Information Interchange – this is a coding system used to represent characters by using numbers.
Backup File – This is a copy of an original used in case the original is corrupted, damaged or lost.
Bit – The smallest unit of data which represents one character. Bits are only ever either '1' or '0'.
Bitmap – An image which is created in a specialised graphics package, such as Microsoft Paint®. This type of file takes up a lot of memory.
Broadband – This is the term given to fast Internet connections which allow the transportation of large amounts of data quickly.
Browser – A piece of software that is used to access web pages.
CD-ROM – A type of storage medium used to hold large amounts of data. It is used for transportation and archiving large files such as images, presentations and large publications. These files cannot be edited as the user will only have write-read access.
Clickable image – any image that when clicked with a mouse causes some form of action to take place.
Compression – Reducing the size of a large file (often used to reduce the size of large image files).
Co-processor – A microchip that handles a specific task. The fact that it is dedicated to a task means the computer can work on other processes.
Crash – A problem arising when the computer stops working as it should.
Cursor – the pointer that is used to represent the mouse on the screen.
Dialogue box – a window that displays options from which the user can make choices.
Dial-up line – a standard phone line used to connect a computer to the Internet. As the name suggests, a connection has to be made each time you wish to use this.
Digital – a series of pulses in discrete levels.
Directory – a structured area that visually represents the files stored in the computer's memory.
Download – to obtain a file from another computer.
DPI – dots per inch – a unit of measurement used to describe the resolution of an image.
DTP – Desk Top Publishing.
Email – electronic mail, used to send messages from one computer user to another.
File – information stored on a disk in a particular, predefined format.
Firewall – a means of preventing unwanted intrusion onto a network which is connected to the Internet.
Flatbed scanner – A scanning device that lies flat on a desk – useful for creating digital versions of text documents or images from printed originals.
Folder – An area in which files or other folders are stored.
Font – a style of text.
FTP – File Transfer Protocol – the means by which files are transferred across a network.
GIF – Graphic Interchange Format – a file format used for images.
Graphical user interface (GUI) – The visual representation of the computer system which it easier for users to run applications and carry out computer-related tasks.
Homepage – the main page of any web site.
Host – the main computer to which users are connected.
Hostname – the name which identifies each computer attached to the Internet.
HTML – Hyper Text Markup Language – this used to show

browsers how to portray web pages on the screen.
Hypertext – a means of linking phrases or images on one webpage to other pages or files simply by clicking on them.
Import – to bring data from one document and place it into another document.
Internet – The world's largest network.
Java – A programming language used to create applications that can run within web pages.
JPEG – Joint Photographic Experts Group – a format used for graphics files, particularly photographs.
kB – kilobyte is a unit of data storage.
LAN – Local Area Network – a network that is situated within a single site or location.
Line art – images which are not photographs shown in only black and white.
Mail merge – Merging data from a database into a document template to create an individual, automated, personalised document.
Mailing list – list of members of a particular email group.
Memory – the chips where data is stored on a computer.
Menu – list of commands.
Menu bar – the bar on which the menus are displayed.
Modem – A device used to connect a computer to the telephone sytem in order to utilise the internet.
MOV – a file extension used to denote a file created in Quicktime.
MPEG – Moving Pictures Expert Group – international standard for video compression. Requires a specialised movie player to be installed on the computer.
Multimedia – a presentation or software that combines a range of media types.
Navigation tools – allows users to find their way around a web site.
Network – a group of computers that are connected in one of a series of ways to form LANs or WANs.
Object-oriented – describes illustrations that are created by mathematical equations.
OCR – Optical Character Recognition – used to scan text from a printed document and convert into digital form so that it can be imported into an application such as a word processor.
On-line – any computer that is connected to another computer.
Operating system – the software which controls the tasks on a computer.
Paste – to insert images or text from another source.
Path – a route used to find files stored on a disk.
PDF – Portable Document Format – used to create documents that can be read on any platform of computer – access to a small plug-in is a all that is required to read the document.
Pixel – Picture element – images are made up of pixels.
Plug-in – small programs that are required to run some web-based applications.
Properties – information about an object or file.
Protocols – the rules governing how data is transmitted between devices.
Quicktime – a file extension for movies.
RAM – Random Access Memory. This is the most common type of computer memory and is used to store programs which are being used at the time.
Resolution – this refers to the sharpness of the images that are being shown.
ROM – Read Only Memory – permanent information is stored here.
Scanner – a device used to convert paper-based originals into digital format.
Search engines – web utilities that are used to help a user find the answer to a query from huge databases of information.
Site – the location of a host.
Spreadsheet – a number-related table which can have calculations applied to it.
SQL – Structured Query Language – the language used by many database systems.

Tags – the formatting codes used in HTML to define the way that content is displayed by a browser.
Taskbar – the bottom of the screen where currently loaded documents and software are displayed.
TIFF – Tag Image File Format – these files are also bitmaps and are their versatility makes them, suitable for transferring files between different applications.
Title bar – the top of a window which contains a name.
Upload – to send a file to another computer.
URL – Uniform Resource Locator – a string of characters that makes up the unique address of a web page.
Web browser – allows you to access HTML documents.
Webpage – the visual representation of HTML codes, text and images.
Word processing – manipulation of text to create documents.
Zipped – compressed version of a program or document.

Index

Index

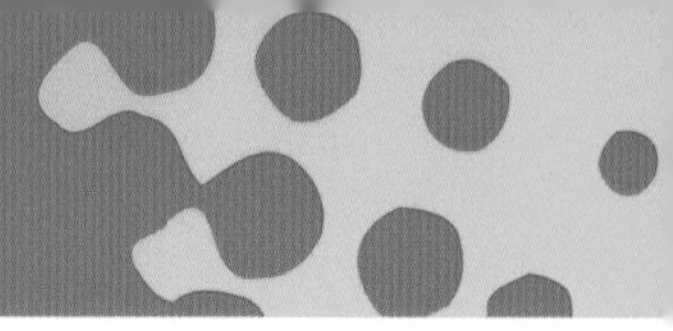

ELECTRONIC END USER SINGLE USE LICENCE AGREEMENT

FOR **DiDA Unit 2: Multimedia CD-ROM Student Version** software published by Hodder and Stoughton Limited (HS) under its Hodder Arnold imprint.

NOTICE TO USER:
THIS IS A CONTRACT. BY INSTALLING THIS SOFTWARE YOU AND OTHERS TO WHOM YOU ALLOW ACCESS TO THE SOFTWARE ACCEPT ALL THE TERMS AND CONDITIONS OF THIS AGREEMENT.

This End User Single Use Licence Agreement accompanies the **DiDA Unit 2: Multimedia CD-ROM Student Version** software (the Software) and shall also apply to any upgrades, modified versions or updates of the Software licensed to you by HS. Please read this Agreement carefully. Upon installing this software you will be asked to accept this Agreement and continue to install or, if you do not wish to accept this Agreement, to decline this Agreement, in which case you will not be able to use the Software.

Upon your acceptance of this Agreement, HS grants to you a non-exclusive, non-transferable licence to install, run and use the Software, subject to the following:

1. Use of the Software. **You may only install a single copy of the Software onto the hard disk or other storage device of only one computer**. If the computer is linked to a local area network then it must be installed in such a way so that the Software cannot be accessed by other computers on the same network. You may make a single back-up copy of the Software (which must be deleted or destroyed on expiry or termination of this Agreement). Except for that single back-up copy, you may not make or distribute any copies of the Software, or use it in any way not specified in this Agreement.

2. Copyright. The Software is owned by HS and its authors and suppliers, and is protected by Copyright Law. Except as stated above, this Agreement does not grant you any intellectual property rights in the Software or in the contents of **DiDA Unit 2: Multimedia CD-ROM** as sold. All moral rights of artists and all other contributors to the Software are hereby asserted.

3. Restrictions. You assume full responsibility for the use of the Software and agree to use the Software legally and responsibly. You agree that you or any other person within or acting on behalf of the purchasing institution shall NOT: use or copy the Software otherwise than as specified in clause 1; transfer, distribute, rent, loan, lease, sub-lease or otherwise deal in the Software or any part of it; alter, adapt, merge, modify or translate the whole or any part of the Software for any purpose; or permit the whole or any part of the Software to be combined with or incorporated in any other product or program. You agree not to reverse engineer, decompile, disassemble or otherwise attempt to discover the source code of the Software. You may not alter or modify the installer program or any other part of the Software or create a new installer for the Software.

4. No Warranty. The Software is being delivered to you AS IS and HS makes no warranty as to its use or performance except that the Software will perform substantially as specified. HS AND ITS AUTHORS AND SUPPLIERS DO NOT AND CANNOT GIVE ANY WARRANTY REGARDING THE PERFORMANCE OR RESULTS YOU MAY OBTAIN BY USING THE SOFTWARE OR ACCOMPANYING OR DERIVED DOCUMENTATION. HS AND ITS AUTHORS AND SUPPLIERS MAKE NO WARRANTIES, EXPRESS OR IMPLIED, AS TO NON-INFRINGEMENT OF THIRD PARTY RIGHTS, THE CONTENT OF THE SOFTWARE, MERCHANTABILITY, OR FITNESS FOR ANY PARTICULAR PURPOSE. IN NO EVENT WILL HS OR ITS AUTHORS OR SUPPLIERS BE LIABLE TO YOU FOR ANY CONSEQUENTIAL, INCIDENTAL, SPECIAL OR OTHER DAMAGES, OR FOR ANY CLAIM BY ANY THIRD PARTY (INCLUDING PERSONS WITH WHOM YOU HAVE USED THE SOFTWARE TO PROVIDE LEARNING SUPPORT) ARISING OUT OF YOUR INSTALLATION OR USE OF THE SOFTWARE.

5. Entire liability. HS's entire liability, and your sole remedy for a breach of the warranty given under Clause 4, is (a) the replacement of the Software not meeting the above limited warranty and which is returned by you within 90 days of purchase; or (b) if HS or its distributors are unable to deliver a replacement copy of the Software you may terminate this Agreement by returning the Software within 90 days of purchase and your money will be refunded. All other liabilities of HS including, without limitation, indirect, consequential and economic loss and loss of profits, together with all warranties, are hereby excluded to the fullest extent permitted by law.

6. Governing Law and General Provisions. This Agreement shall be governed by the laws of England and any actions arising shall be brought before the courts of England. If any part of this Agreement is found void and unenforceable, it will not affect the validity of the balance of the Agreement, which shall remain wholly valid and enforceable according to its terms. All rights not specifically licensed to you under this Agreement are reserved to HS. This Agreement shall automatically terminate upon failure by you to comply with its terms. This Agreement is the entire and only agreement between the parties relating to its subject matter. It supersedes any and all previous agreements and understandings (whether written or oral) relating to its subject matter and may only be amended in writing, signed on behalf of both parties.